DIRECTIONS IN DEVELOPMENT

Reengaging in Agricultural Water Management
Challenges and Options

Challenges and Options

Reengaging in Agricultural Water Management

Challenges and Options

THE WORLD BANK
Washington, DC

Cover photo: Tomas Sennett/World Bank

ISBN-10: 0-8213-6498-7
e-ISBN: 0-8213-6499-5
ISBN-13: 978-0-8213-6498-7
DOI: 10.1596/978-0-8213-6498-7

Library of Congress Cataloging-in-Publication Data has been applied for.

Contents

Boxes

Foreword

Results of past investments in agricultural water management have been mixed. On one hand, irrigated agriculture expansion and productivity increases have made a major contribution to meeting fast-rising world food demand, supporting rural economic development, and reducing poverty. In the past 40 years, demand for food in the developing countries has gone up sharply. Production and average yields of irrigated crops in these countries have responded to this demand by increasing two- to fourfold. On the other hand, water availability for agriculture is increasingly constrained; environmental stresses are growing as many river basins approach the limits of water and land resources; the pace of irrigation expansion is slowing down; and drainage continues to be neglected. Furthermore, rainfed areas where most poor people live have been largely bypassed by public investment in water management.

The strong demographic and increased income push to food demand is expected to continue and it is anticipated that irrigated agriculture will provide close to 60 percent of the extra food needed over the next 25 years. In addition to more intensified and diversified irrigation and to some expansion of the irrigated area especially in Sub-Saharan Africa, a major effort is needed to enhance water management in rainfed agriculture. The role of Government is changing, responsibility is decentralizing, farmers are playing a greater role in decisions and investments, and more and more, markets are driving growth. New solutions are needed, based on improved management options and widely available technologies.

How to meet ever-rising demand for food while at the same time increasing farmer incomes, reducing poverty, and protecting the environment, all from an increasingly constrained water resource base, are the main challenges facing agricultural water management.

This report describes what governments, international agencies, rural people, the private sector, and others can do to ensure that these challenges can be met. It sets out the changing context of demand and supply for agricultural water and identifies the policy, institutional, and incentive reform options that help meet these challenges. It articulates priorities for investment, indicates options for adjusting the respective roles of the public sector and other stakeholders, and sets out how agricultural water management can best integrate upstream into water resources management, and downstream as an input provider into the agricultural economy.

Sushma Ganguly   *Kevin Cleaver*
Sector Manager   *Director*
Agricultural and Rural Development   *Agricultural and Rural Development*

Acknowledgments

Preparation of this report has been managed by the Agricultural and Rural Development Department. Christopher Ward, Salah Darghouth, Gayane Minasyan, and Gretel Gambarelli wrote the report. They would like to extend their sincere thanks and appreciation to Ariel Dinar, Safwat Abdel-Dayem, Shobha Shetty, Ijsbrand de Jong, Claudia Sadoff, Douglas Olson, Geoffrey Spencer, Julia Bucknall, Imane Akalay, Musa Asad, Gabriel Azevedo, and Jean-Christophe Carret for their input and feedback throughout the preparation of the report.

The authors also wish to acknowledge constructive comments from the peer reviewers: Joseph Goldberg, Ashok Subramanian, Keith Pitman, John McIntire, Rory O'Sullivan, Herve Plusquellec, Jean-Marc Faures (FAO), David Molden (IWMI), and Mark Rosegrant (IFPRI), as well as from Thierry Facon (FAO), Manuel Contijoch, Vijay Jaganathan, and Peter Koenig. Many others provided support, including several members of the World Bank Water for Food Team as well as Melissa Williams, Jonathan Agwe, Marisa Baldwin, Corazon Solomon, and Immaculate Bampadde.

Kevin Cleaver and Sushma Ganguly have provided invaluable support throughout the development of this report. Their contributions have lent clarity and power to a document that is intended to support the development community in its efforts to reengage in reformed agricultural water management that contributes to economic growth, poverty alleviation, and environment protection.

Salah Darghouth
Water Advisor
Agriculture and Water Development

Acronyms and Abbreviations

ADID	Agricultural Directions in Development Report
ARD	Agricultural and Rural Development Department, World Bank
AWI	Agricultural water investment (sourcebook)
AWM	agricultural water management
BCM	Billion cubic meter(s)
BOO	build-operate-own
BOT	build-operate-transfer
CACG	Compagnie d'aménagement (Regional water company) des Côteaux de Gascogne, France
CAS	Country Assistance Strategy
CDD	community-driven development
CGIAR	Consultative Group on International Agricultural Research
CWRAS	Country Water Resources Assistance Strategy
DID	Directions in Development
DRAINFRAME	Drainage Integrated Analytical Framework
ECA	Europe and Central Asia Region, World Bank
EIA	environmental impact assessment
EMP	Environmental Management Plan
ET	evapotranspiration
EU	European Union
FAO	Food and Agriculture Organization
FEWS-NET	Famine Early Warning Systems Network, USAID
GDP	gross domestic product
GW-MATE	Groundwater Management Advisory Team
I&D	irrigation and drainage
ICARDA	International Center for Agricultural Research in the Dry Areas
ICID	International Commission on Irrigation and Drainage
ICR	Implementation Completion Report

ICRISAT	International Crops Research Institute for the Semi-Arid Tropics
IDE	International Development Enterprises
IFAD	International Fund for Agricultural Development
IFC	International Finance Corporation
IFPRI	International Food Policy Research Institute
IN	Investment Note
INPIM	International Network for Participatory Irrigation Management
IP	Innovation Profile
IPM	integrated pest management
IPTRID	International Programme for Technology and Research in I&D
IRRI	International Rice Research Institute
IWMI	International Water Management Institute
IWRM	Integrated Water Resources Management
LAC	Latin America and the Caribbean Region, World Bank
LSI	large-scale irrigation
M&I	municipal and industrial
MCM	million cubic meter(s)
MDG	Millennium Development Goal
MENA	Middle East and North Africa Region, World Bank
NAFTA	North American Free Trade Agreement
NENA	Near East and North Africa Region (including MENA, plus Afghanistan, Turkey, and Cyprus as classified by FAO)
NGO	nongovernmental organization
NIA	National Irrigation Authority of the Philippines
O&M	operation and maintenance
OECD	Organisation for Economic Co-operation and Development
OED	Operations Evaluation Department (World Bank)
PAD	Project Appraisal Document
PHM	Participatory Hydrological Monitoring, Andhra Pradesh, India
PIP	private irrigation promotion
PPP	public-private partnership
PRSP	Poverty Reduction Strategy Paper
RAP	Rapid Appraisal Procedure
SAED	Societe d'Amenagement des Eaux du Delta du Fleuve Senegal
SHI	smallholder irrigation
SSA	Sub-Saharan Africa

UN	United Nations
WARDA	African Rice Center (*Centre du riz pour l'Afrique*)
WBI	World Bank Institute
WHO	World Health Organization
WRSS	Water Resources Sector Strategy, World Bank
WTO	World Trade Organization
WUA	water user association
ZATAC	Zambia Agribusiness Technical Assistance Center

Executive Summary

In recent years, agricultural water has helped meet rapidly rising demand for food, and has contributed to the growth of farm profitability and poverty reduction as well as to regional development and environmental protection.

After several decades of publicly funded surface irrigation, and more recently of privately developed groundwater irrigation, remaining opportunities to harness new resources for agriculture are fewer and more expensive. Investment is increasingly focused on rehabilitating and improving the existing systems. However, water productivity remains generally low, and returns to public investment generally disappointing, especially in large-scale irrigation. New solutions are needed, based on new management options and widely available technologies. The role of government is changing, responsibility is being decentralized, farmers are playing an increasingly important role in decisions and investment, and more and more, markets are driving growth. How to meet ever rising demand for food while at the same time increasing farmer incomes, reducing poverty, and protecting the environment, all from an increasingly constrained water resource base, is the main challenge facing agricultural water management (AWM).

The Bank's recent corporate strategies for Rural Development, Water, and Environment, edited by the Environmentally and Socially Sustainable Development Network, call for a reengagement in AWM, and provide general principles on how that could happen. Most growth should come from improvements in water productivity; sustainable increases in farmer incomes are essential, with a focus on the poor; institutional improvements are needed to increase efficiency of resource use; and water for agriculture has to be used sustainably within an integrated approach.

The overall goal of this report is to give strategic focus to implementation of the AWM components of the corporate strategies. Its specific objectives are to set out the changing context of demand and supply for agricultural water; to identify the policy, institutional, and incentive reform options that will accelerate productivity improvements and pro-poor growth; and to articulate priorities for investment in AWM. It is also intended to define the role of the public sector and other stakeholders, and to set out how AWM can be best integrated upstream into water resources management, and downstream as an input provider into the agricultural economy.

The primary audiences for the paper are policy makers and project managers in our partner countries and development organizations, as well as World Bank country and sector managers and task team leaders. There is broad interest among these partners in collaborating with the Bank on developing a new AWM agenda. The implications of this report are many and far-reaching. It will be used (a) as a platform for a wide global dissemination and consultation with our partners on the best policy, institutional, and investment options to reengage in AWM; and (b) as the strategic framework for the preparation by the Bank of its action plan in the sector.

THE ACHIEVEMENTS AND CHALLENGES OF AGRICULTURAL WATER MANAGEMENT

AWM is diverse and has strong links to other sectors and to the broader economy. AWM is not a goal in itself but part of a process of resource management that provides a key input to agricultural production and farmer incomes. It includes irrigation and drainage, water management in rainfed agriculture, recycled water reuse, water and land conservation, and watershed management. It covers all irrigated agriculture, whether fed by surface water or groundwater, including both public schemes and millions of private individually irrigated farms, in a wide range of agro climatic conditions, and in a broad set of production systems and water management contexts. AWM is at the crossroads between four areas of public policy for sustainable growth: water resource management, agriculture, rural development, and the environment. AWM also interacts closely with broader aspects of macroeconomic policy for growth.

Irrigated agriculture has been vital to meeting quickly rising food demand. In the last 40 years, developing country demand for food has tripled, increasing much faster than population growth rates, as nutrition has improved. Food production in the developing world has almost kept pace, with an enormous rise in production (up two-and-a-half times during this period). Crops that are mostly irrigated—such as rice, wheat, maize, and cotton—saw production increasing since the early 1960s two- to fourfold. The production of irrigated fresh fruit and vegetables increased particularly quickly over the period—by four to six times, and these crops now account for over one-fifth of all developing country agricultural exports. Two-thirds of the increase in crop production has come from yield increases, rather than from expansion of the cropped area (except in Sub-Saharan Africa). Average yields of rice and maize more than doubled, and wheat yields went up threefold.

Irrigation continues to expand but now the pace is slowing. For developing countries as a whole, the irrigated area more than doubled over the last 40 years, and by 2000 covered 234 million hectares (ha) (representing 85 percent of the world's total irrigated area of 276 million ha in 2000), about half the land estimated by FAO to be potentially irrigable. However, the pace of development has now slowed significantly: annual rates of expansion of around 2 percent a year in the 1960s and 1970s slowed to hardly 1 percent in the 1990s. Many countries now face constraints to expansion, particularly from social and environmental concerns. The low productivity of many existing schemes has prompted a change in investment policy in the sector, away from new infrastructure and toward programs that improve the performance of existing schemes.

Water availability for irrigation is increasingly constrained. Irrigation accounts for 85 percent of water withdrawals in developing countries, and the rapid growth of the sector has been based on the availability of these huge quantities of low-cost water. Now rising demand for agricultural water faces increased demand from domestic and industrial uses. Many areas are already enduring competition for water and rising marginal costs. For years, groundwater provided a profitable new resource, but in many basins groundwater is now being mined rapidly.

Governments have led the expansion of large-scale irrigation but performance has been suboptimal. With strong investment and management input from governments, large-scale irrigation has contributed to rapid increases in food production, the major public policy goal. However, the supply-led approaches and large-scale irrigation infrastructure that were to fuel growth have resulted in bureaucratic institutions that lack the structure and incentives for efficient management, and in inflexible water delivery systems not capable of responding to farmer needs.

Water productivity has shot up but there is massive room for improvement. The increase in water productivity in recent years has been spectacular: over the period 1961–2003 the water needed to produce food for one person halved from six cubic meters a day to less than three cubic meters a day. Over the same period, the production of rice and wheat went up by 100 percent and 160 percent, respectively, but with no increase in water use. However, in many basins, water productivity remains startlingly low and takeup of modern technology is slow: drip technology has been adopted on less than 1 per cent of irrigated lands worldwide.

AWM has contributed to poverty reduction in irrigated agriculture, but improvements have largely bypassed farmers in the rainfed areas. AWM has made a substantial contribution to poverty reduction, although irrigation development has not often targeted the poor specifically. The groundwater revolution also has a significant poverty-reduction impact, bringing a reliable water source right onto the farms of poor people. However, the rainfed areas where most poor people live have been largely bypassed by the Green Revolution and by public investment in enhanced water management.

Environmental and social impacts of irrigation have been positive and negative, but stresses are growing. As water and land managers, farmers are also stewards of the environment, and they provide many environmental services and amenities to society. AWM and its infrastructure help mitigate the impacts of drought and floods, stabilize river flows, and reduce erosion and silt loads. They have contributed to shaping the countryside and to social and cultural values. However, tension between agricultural production and protection of natural resources is growing. Farmers face increasing difficulty in fulfilling their trusteeship role as many countries approach the limits of water and land resources. Much irrigated land suffers from drainage problems—about half a million ha go out of production each year. The third-party environmental costs and risks of irrigated agriculture have grown: loss of environmental water flows; groundwater overexploitation; pollution; destruction of natural habitats and livelihoods through drainage of wetlands; and waterborne diseases.

Overall, there have been significant advances in AWM but challenges are great, especially in Sub-Saharan Africa. Overall, the pattern of recent years has been of significant advances in AWM and in productivity, making a major contribution to farmers' incomes, poverty reduction, and regional and national development. The challenges ahead are, however, enormous, and nowhere more so than in Sub-Saharan Africa, where per capita cereals consumption is only half that of East Asia and where one-third of farmers remain hungry.

More details are included in chapters 1 and 2.

THE CHANGING GLOBAL AND NATIONAL CONTEXTS FOR AWM

The global debate on water resources management and food security is sharpening the agenda for AWM. Water resources and food production are increasingly global issues, and now debate is beginning to focus attention on key AWM questions, such as the potential conflicts between water for food and water for nature; the environmental impacts of irrigation intensification; and the trade-offs between low food prices and producer incentives and incomes. Pioneering work by the International Food Policy Research Institute (IFPRI), the International Water Management Institute (IWMI), and the Food and Agriculture Organization (FAO) has started to bring the issues to the fore, with major recent publications by these agencies exploring the water-for-food challenge. International research is now starting to reflect the growing emphasis on water productivity.

Changes in the global trade environment and national marketing strategies are of critical importance. The irrigation sector depends on market-derived incentives for its future, and some countries—with rapidly growing economies—have begun to move from a supply-driven food production strategy toward market-driven policies for AWM that focus on productivity and incomes. However, constraints to market-driven approaches persist: remaining trade barriers are predominantly on irrigated agricultural products (such as rice, wheat, cotton, and sugar), and access to national and international markets for smallholders is constrained by domestic restrictions on market development and by the lack of organized smallholder supply chains. Where access does exist, as for horticultural products, the dynamic impact of market-driven growth on irrigation development and productivity has been great.

Water resources management is changing, and environmental and social concerns are growing. Responses to growing scarcity, to increased competition among sectors, and to growing environmental and social concerns include integrated and basin management approaches and demand management measures. On the supply side, there are fewer new diversion and storage projects, and more consideration of reuse of wastewater and drainage water. Climate change is increasing the existing vulnerability of farmers. Investment policies are starting to move toward upgrading and management improvements, although very slowly. Consideration of the environmental and social impacts is becoming an important factor in AWM, with broader understanding of the multifunctionality of water and of human and ecosystem interactions. Environmental and social concerns are increasingly mainstreamed.

The roles of the respective stakeholders are changing. The role of government in AWM has begun to change, with tentative moves toward a greater role for users. There has been some decentralization, and the participatory irrigation management movement has caught on in more than 50 countries. However, few public irrigation schemes have become financially self-sustaining, and cost recovery generally remains low. Investment by farmers and other private sector investment is substantial, particularly in small-scale irrigation and private groundwater irrigation, which alone account for over half of the irrigated area worldwide. There are some initiatives in public-private partnership (PPP) or large-scale irrigation, but they remain very timid.

More details are included in chapter 3.

THE FUTURE STRESSES AND RISKS CAUSED BY RISING FOOD DEMAND AND INTENSIFICATION OF IRRIGATED AGRICULTURE

The strong demographic push to food demand is expected to continue. For the developing world as a whole, population is projected to increase by half over the period 1999–2030. Developing countries' food self-sufficiency ratio is expected to decline from 91 percent to 86 percent, and their food trade balance is expected to turn sharply negative (US$50 billion annually by 2030). Nations with fast-growing economies will be able to import an increasing share of their basic food needs, which will stimulate investment in higher-value irrigated agriculture where markets exist. The poorer nations, particularly in Sub-Saharan Africa, are likely to focus on strategies to develop irrigated agriculture where investment costs are not too high, and to improve food crop production in currently subsistence agriculture environments. AWM will be an essential element in both strategies.

Intensified irrigated agriculture will provide more than half of the extra food. FAO has estimated that crop production in developing countries needs to increase at about 1.6 percent per year over the next three decades—a demanding challenge, although only half the rate of growth recorded in the last 10 years. Projections by FAO and IFPRI/IWMI are that irrigated areas are likely to have to provide more than half of this increased production. As water and land resources are constrained, further water productivity improvements will be essential. Water productivity improvements in large-scale irrigation are possible, but require major programs of "modernization"—a combination of institutional change and investment in system improvement. There is scope, too, for groundwater productivity to improve. In addition to technical choices, farmers have multiple choices to increase income from their production, particularly through diversification into production of fruits and vegetables and other higher-value irrigated crops.

However, over 40 percent of the extra food will have to come from intensified rainfed farming in coming years, for which improved water management is essential. Rainfed cereals yields would need to increase—IFPRI/IWMI (2002) estimates by more than 40 percent by 2025. The water productivity challenge in rainfed farming is how to introduce accessible technical solutions without increasing risks. Known techniques for soil moisture conservation and water harvesting—and some technologies for rainfed areas such as low-cost supplemental irrigation—can have high returns.

Growing water scarcity will have to be managed. In most parts of the world, the water available to irrigation will be constrained further, and irrigation consumption will grow much more slowly than consumption in municipal and industrial uses. In Asia overall, IFPRI/IWMI (2002) projects that water consumption by all users will increase by 14 percent by 2025, but irrigation consumption will go up by only 1 percent—and in water-constrained China, irrigation consumption is even projected to decline. Water stress will create a strong push to improve water productivity and to strengthen the use of demand management approaches. In many river basins, intersectoral competition will be a critical problem. Increased withdrawals for irrigation will be limited, and mechanisms for allocating water equitably between sectors will be needed. Groundwater depletion from increased irrigation will continue and may accelerate. Governments and users will have strong incentives to work on reducing rates of depletion.

There is some potential for expansion of the irrigated area. FAO estimates that the irrigated area in developing countries could increase by almost 20 percent (40 million ha) in the period 1997–9 to 2030. Some increase in the irrigated area will be supplied by diversion from structures already in place. Elsewhere, some new water withdrawal projects for irrigation would be undertaken. In Sub-Saharan Africa and Latin America in particular, there is technically scope for expansion of irrigation.

Risks for the environment and society will increase. As irrigated agriculture is intensified and as additional irrigation capacity is developed and drawdown of groundwater continues, environmental risks will increase. It will be essential to manage these risks using the technical, managerial, and economic instruments that have been developed progressively in recent years.
    More details are included in chapter 4.

POLICIES, INSTITUTIONS, AND INVESTMENTS TO PROMOTE AGRICULTURAL WATER IN DEVELOPMENT

This section summarizes the options and trade-offs for improving AWM, beginning with the farmer's perspective and then treating in turn options at the system or area level, at the sectoral level, at the level of the nation and the macroeconomy, and finally in the global context. These reform options are described in detail in chapters 5 and 6 of the report. Relevant sections of those chapters are noted by section number.

The farmer's perspective

The farmer's main objectives are to increase his or her income and assets sustainably and to reduce vulnerability. Water security—access to assured water supplies—is an essential prerequisite. The farmer thus needs to have a say in the management of the irrigation system, which will provide a water service of quality as well as a secure water entitlement. These interests set the AWM reform priorities: irrigation modernization, user participation, water rights, and demand-driven investment. For profitable farming, the farmer also requires access to efficient input and output markets, and to cost-effective technology. These needs set the priorities for agricultural policy: market development, and research and technology transfer. How the farmer's interests and needs in AWM can be met is described in detail in the report and summarized below.

Options at the system or area level

"Modernizing" large-scale irrigation. In large-scale irrigation (LSI), the objective is to improve farming profitability sustainably through improved service at the least public cost. The inflexible water delivery systems and bureaucratic institutional design that characterize much LSI make response to changing markets and profit opportunities difficult. Further improvements in profitability have to be made through integrated system modernization, that is, by turning both the irrigation delivery system and the institutional structure around to focus on delivering a sustainable, efficient, and demand-responsive water delivery service. LSI modernization thus requires an integrated package of physical improvements and institutional change in addition to agronomic improvements.

Physical improvements will include a broad range of "hardware" investments and related management practices to assure an efficient, least-cost water service delivery that meets farmer needs. Optimization tools have been developed that allow the most cost-effective investments to be selected. (Section 6.1)

The parallel *institutional changes* to create a demand-responsive water service delivery typically include a reduction in the role of governments in management and financing, and promotion of decentralization, agency accountability, and scheme financial autonomy as an interim milestone toward full scheme management transfer. Efficiency improvements should be introduced to reduce costs and expand the revenue base: in the irrigation reform in Victoria, Australia, 80 percent of the improvement in financial performance came from system efficiency gains and an expanded revenue base, and only 20 percent from increased water charges. Water user associations have proved effective in modernization programs, and user participation should be included at each step of the decisionmaking process. Scaling up to water boards or user federations should be encouraged. Irrigation management transfer should be undertaken when the conditions are right and should generally be a carefully designed and implemented, medium- to long-term process. A possible complement is to involve the private sector through public-private partnerships (PPP). PPP brings in a "third professional party" that can be the catalyst for improved management and the genesis of a corporate culture. (5.5)

A vital component of institutional change—scheme financial autonomy—depends on cost recovery. Low cost recovery leads inexorably to poor service, and covering scheme costs is a mandatory objective: if systems are to deliver quality service, somebody has to pay for it. If irrigators cannot pay, then government must. Globally, this is an area where scant progress has been made to date, and more work is needed. There should be global dialogue to establish internationally valid benchmarks and targets. Within a scheme, it has to be clear what investment, operations and maintenance, and other costs should be recovered from whom, and how—for example, the costs of upstream works could be financed by government, downstream works at the tertiary and quaternary level by the irrigators, with cost sharing for the secondary canal level. (5.4)

Overall, irrigation "modernization" is a process implemented over an often lengthy period, with changes sequenced and integrated as needed. Priorities are a focus on the objective of farmer profitability through improved service delivery; a market-driven demand orientation; integration of physical investment, agronomic improvements, and institutional change including a reduced role for government; involvement of users throughout; efficiency improvements to reduce costs; and scheme financial and managerial autonomy. (6.1)

Improving the profitability of small- and medium-scale irrigation. Water productivity on traditional and small-scale AWM systems is typically low. Government support is best provided through community-driven approaches and financing mechanisms, or working through nongovernmental organizations (NGOs) as part of a broader package of rural development that ensures that rural and market infrastructure develop in step with one another. Participatory irrigation management (PIM) and irrigation management transfer (IMT) should be systematically encouraged. An element of matching grants will be necessary. The agenda should include research for the development of affordable irrigation technologies. New approaches use the market to develop appropriate technologies and to disseminate them. (6.1)

Ensuring more sustainable development of groundwater irrigation. Unplanned mining of groundwater has severe costs for the rural economy, particularly for the poor, and the challenge is to recover sufficient control to allow optimum economic benefits to be achieved. First best solutions rely on a rights-and-regulation framework, but in most countries this will be a very long-term solution. The alternative is to strengthen existing rights and promote self-regulation, with supporting changes to the incentive framework. In particular, governments need to eliminate energy subsidies, which drive overdrafts everywhere. Demand-side measures to improve the efficiency of water use should be combined with supply-side measures, such as aquifer recharge enhancement, rainwater harvesting, drainage, and urban waste-water reuse. With a handful of possible exceptions, such as Jordan, no developing country has succeeded in recovering control over groundwater, and prospects for eliminating overdraft completely are limited. However, using the institutional, economic, and technical tools discussed, countries may move toward more "planned" depletion, where a slower pace of mining may allow a less water-intensive economy to develop without severe shock or negative impacts on the poor. (5.2)

Enhancing water productivity in rainfed agriculture. Improving water availability and productivity in rainfed agriculture and watersheds is essential for household food security and poverty reduction, yet solutions are much less evident than for irrigated areas. There is a significant research agenda, particularly on land and water management and agronomic practices, but priorities are the transfer of existing technology, the development of market outlets, and physical investment in rural infrastructure and in water control structures. Market-driven integrated approaches that reduce risk, and that involve community participation throughout are most likely to succeed. (6.1)

Developing and integrating sector policies for AWM

At the sectoral level, policies for water resources management, agriculture, rural development, and the environment need to mesh to support sustainable, market-driven growth in rural incomes based on improved AWM.

Water resource management policies. Critical areas where water resources management and AWM need to interact are basin planning, incentives to water productivity, nonconventional water, and water rights.

AWM has to be treated within an *integrated water resource management framework* in which basin plans aim at accommodating often conflicting objectives such as economic efficiency of water allocation, equitable water distribution, and environmental protection, including drainage needs and environmental flows. The basin approach allows the productivity of agricultural water to be managed by reducing the amount of water depleted from the water balance: measurement of returns per unit of water lost through evapotranspiration should become the yardstick of productivity. (5.2)

As water scarcity increases, the whole *incentive structure* has to promote water productivity. Demand management should combine the price signals that result from the macroeconomic, trade, and fiscal regimes and from agricultural and irrigation sector policy with nonprice factors such as rationing, asset transfer, or cost sharing on investment to create incentives to water productivity. (5.4)

Particularly in water-scarce countries, investment in *reuse of treated waste and drainage water* can offset water scarcity, but there are trade-offs that need to be managed in an overall basin context, including human health risks, pollution, and reduced environmental flows. Reuse of wastewater is a key area for investment. Governments have to determine reuse policies and establish the regulatory framework, but users should be partners in the development of programs. (6.1)

Established *water rights*—especially tradable rights—should improve water productivity and promote investment. However, on large schemes where quantities are uncertain and service delivery weak, attribution of legal rights is hard, and development of firm entitlements, often at the group level, should form part of the modernization programs. Some countries—Jordan, for example—have introduced formal rights by developing over time a flexible legal framework of entitlement and transfer, with capacity building. They have also formalized existing informal markets. (5.2)

Agricultural policies. Three areas of agricultural policy are of critical importance for AWM: market development policies, food policy, and policy for technology development and transfer.

Development of internal and export markets is the most important driver of farm profitability, together with efficient allocation of agricultural water, increased water use productivity, and investment and modernization in irrigated agriculture. Domestic market reforms—liberalization, privatization, subsidy removal—should complement external trade reforms and create an enabling environment for irrigated production, which encourages inward and domestic investment and provides for secure contractual arrangements. Development of exports in horticulture, for example, may require governments to take an active role in developing the behind the border agenda in trade facilitation. Government's role is best undertaken in partnership with the private sector. In addition, strategic investment to promote markets and create market and transport infrastructure can be critical to the development of irrigated agriculture. Development of markets and roads in the Nigerian *fadama* combined with access to groundwater boosted profitability by three times and more. (5.3)

Food policy has driven much public investment in irrigation, successfully supplying cheap food but often keeping irrigators poor and reducing investment returns. Food security can best be increased by channeling scarce water to the most profitable enterprises of the poor, not by targeting food production per se. The emphasis has to be on efficient resource allocation and on the development of markets to add value to the production of the poor and to ensure that food is available. For poorer countries, escaping from the poverty trap requires taking some risks in moving toward a market-driven irrigated agriculture. Better off countries (including China) should consider moving progressively away from strict self-sufficiency goals to high-value irrigated production. Where food policy changes, support and safety net programs may be needed. (5.3)

Technology development and transfer is essential to growth—but it has to be market-driven. Considerable AWM technology is available, but farmer adoption has been slow. Currently, just 3 percent of the irrigated area worldwide uses pressurized irrigation, and the scope for expansion is enormous. Technology adoption is best promoted by encouraging the development of profitable product markets. Governments should also work with the private sector to develop technology and promote its adoption through the market. (6.1)

Rural development policies. Rural development policies target sustainable improvements in livelihoods. Irrigation helps reduce *poverty* through increased food output, higher demand for employment, and higher real incomes, and also drives a local multiplier effect to increase nonfarm rural output and employment. Irrigation also reduces vulnerability by stabilizing output, employment, and income. In general, irrigation has the most poverty-reducing impact where (a) there is equity in land distribution; (b) investments and water charges are designed with the needs of the poor in mind; (c) schemes are well managed and provide good water service; and (d) users are involved in management. There may, however, be negative impacts on the poor, and irrigation is not always the most efficient pro-poor investment available. Policy analysis and poverty-reduction strategy papers should explicitly examine poverty-reduction aspects of AWM, and poverty reduction should be built into AWM investment programs. Programs should give priority to (a) pro-poor rainfed agricultural water (and land) investment; (b) low-cost irrigation technologies, preferably through the market; (c) use of community-driven development and social fund approaches to AWM investment; (d) small-scale irrigation and water conservation investments; (e) targeting large-scale irrigation investments toward pro-poor entry points; and (f) diversification into higher-value irrigated crops. Care has to be taken to ensure that the benefits of public support go principally to the poor. (5.7, 6.4)

Women are stakeholders in AWM—and a poverty target group—yet they are widely disregarded in policy and programs. Women should be systematically involved in AWM projects, and economic and social analysis and mechanisms of participation and inclusion should be adapted to increase the effectiveness of women's participation. (5.6)

Environmental policies. The considerable global experience on *managing environmental risks* needs to be applied both to intensification and to expansion of irrigation. At the macroeconomic level, the main instrument to guide farmers to environmentally friendly practices should be the incentive structure, which should reflect the value of environmental goods, services, and costs, for example, reducing energy subsidies or making cost-sharing grants for terrace maintenance. At the sectoral level, environmental concerns need to be mainstreamed into all aspects of water management and agricultural policy, including into research and technological innovation and adoption in AWM. Expansion of irrigation should take place within basin plans, using safeguard approaches. Particular attention should be given to the protection of environmental flows and of groundwater resources. (5.8)

Much of the world's irrigated lands suffer from *drainage* problems, and an estimated 20–30 million ha need improved drainage. Developing countries should allocate more resources to drainage investments within an integrated water resources management framework, using participatory approaches and planning tools to take account of the social, economic, and technological aspects. (6.1)

Policy integration. Improving the profitability of irrigated farming requires a combination of actions at the farm, scheme, and sectoral levels. The common thread at all levels is that of market-driven incentives, but a wide range of policies, institutional reforms, and investments is needed to steer irrigated agriculture onto a sustainable growth path. These measures will vary according to local conditions, and sequencing and prioritization need consideration. In the 1970s and 1980s, investment in large-scale irrigation in Morocco and Jordan created irrigation networks capable of better service, but the scheme-level improvements in institutions, the links to external markets, and the integration of scheme water use within efficient basin plans has come much more slowly, so that water productivity and farming profitability are only now improving. The integration of scheme and sectoral measures into the broader framework of national macroeconomic policy for growth is also key to driving productivity and profitability.

Macroeconomic policy and AWM

At the macroeconomic level, the objective is national economic growth through efficient resource allocation. At this level, the roles of government and other stakeholders are determined, and fiscal policies on budget support and investment are decided. The political economy of vested interests and competing objectives is also important at this level.

Governments should be responsible for *core public sector tasks* related to AWM: integrated water resources management, environmental protection, research and technology transfer, and rural infrastructure. In addition, governments should correct market failure through interventions in poverty reduction, water pricing, and the development of product and financial markets. Beyond these functions, governments should seek broader engagement of other stakeholders—farmers, NGOs, and the private sector—in a process of decentralization and inclusion. (5.5)

Government budget support finances about half of the US$30–35 billion invested globally in irrigation each year. Past patterns of support have generally reduced the cost of water, giving little incentive for efficient use and creating distortions in the market. Budget support should be realigned with policy goals such as water use productivity, farming profitability, and poverty reduction—for example, the current generation of "smart" cost-sharing subsidies on drip irrigation. (5.4)

Public investment in AWM should be guided by the lessons of experience: integration within basin plans, decentralized management, participatory approaches, and financial sustainability at least cost to the government budget. Some irrigation expansion and new water resource withdrawals will be justified by rising demand for agricultural products in coming years. Projects will have to be justified in terms of their impacts at the overall basin hydrological and welfare level in a way that is seen to be fair to all stakeholders. Environmental and social risks will need systematic mitigation, too. That expansion of large-scale irrigation is the best investment available in AWM will need to be demonstrated, because returns to other AWM investments may be much higher. However, investment in new infrastructure is certainly justified in countries such as Ethiopia, which has abundant water and a level of storage infrastructure per capita less than 1 percent of that of North America. The private sector should be involved through PPP wherever possible. Where feasible, irrigation expansion projects should be integrated into multipurpose programs to ensure inclusion in the integrated water management framework and to improve the economics of the irrigation component. Water infrastructure should increasingly be seen as a means of increasing water security by reducing vulnerability to exogenous shocks such as floods, drought, and hydrological variability. (5.1, 5.5, 6.1)

Returns to investment in AWM are often higher than has been estimated in recent years, and this will increase the attractiveness of investment. Benefits from multifunctionality of irrigation and drainage investments and from the multiplier effect of direct and indirect job and wealth creation in the economy have been understated. One study in Pakistan found that total benefits of irrigation were 12 times the direct, on-site benefits when all quantifiable economic and social benefits were accounted for. Climate change and hydrological variability cost the Ethiopian economy over one-third of its growth potential, and returns to irrigation are correspondingly high. Where profitable markets are available, economic returns to agricultural water are competitive. (6.2)

Policies, institutions, and resource allocation are shaped in part by the *political economy* of each nation. The structure of established interests means that in any change there will be losers as well as winners, and reforms in AWM typically have high political transactions costs. Governments need to enlist support for reforms through transparent and inclusive processes. Reforms need champions, and should be piloted to show how benefits outweigh costs. Incentives need to be built in, including early benefits for "winners" and support measures for "losers." (5.5)

AWM in the global context

At the global level, three major issues will affect AWM: trade reform, climate change, and the global research agenda.

Trade reform policies will strongly influence water productivity and profitability in agriculture by opening up external markets. The impacts of trade reform on irrigated agriculture should be carefully assessed before reforms are undertaken, because impacts on the irrigation economy can be negative as well as positive. A phased program including economic mechanisms and social support programs should be developed to help the adjustment toward free trade. Nations should invest in institutions and technology, because trade-driven growth requires a knowledge-intensive irrigated agriculture. (5.1)

Climate change creates greater risks and uncertainties, which should be dealt with by a risk management approach. At the strategy and policy level, adaptation to climate change needs to be factored into economy-wide modeling and poverty-reduction strategies. Increased hydrological variability will drive changes in investment programs, because investments in water storage and water productivity will become more profitable in many areas.

Research and technology transfer are vital to obtaining productivity improvements in AWM. Technical research priorities should focus increasingly on water productivity and on AWM for rainfed farming. Institutional, social, and economic research will also be vital on such aspects as large-scale irrigation management and modernization, and poverty-reduction impacts. Research institutions and governments should forge partnerships with the private sector, which is already very active in development and dissemination of irrigation technology. (5.1)

THE PRIMARY MESSAGES OF THE REPORT:
TOWARD AN ACTION PLAN

The report sets AWM as an input to farming and as a key factor in farmer incomes, in agricultural growth and exports, and in poverty reduction. This economic context defines two underlying themes: an emphasis on productivity of water use and the need for market-driven approaches. These themes have driven the key messages of the report:

- The setting of AWM within an integrated water resource management context, ensuring both efficiency in allocation of water between sectors and the integration of the productivity of agricultural water within the broader context of evapotranspiration from the hydraulic system.
- A focus on ways to increase water productivity and farming profitability through markets and the incentive structure, through investment, and through technology development and adoption.
- A move toward new institutional arrangements, which give more responsibility and say to farmers, engage the energy of the private sector, and reduce government's role.
- An emphasis on integration of policies, institutional change, and investments to achieve efficient outcomes in all aspects of AWM from modernization of large-scale integration to enhancing water management in rainfed agriculture, and on the sequencing and prioritization of change processes.
- A pragmatic approach to intensification and expansion of AWM, using participatory approaches and new methodologies to make sure that social and environmental concerns enhance the economics and sustainability of investments, and ensuring that the broader benefits of AWM are captured.
- Increased attention to the potential for reducing poverty, and the systematic factoring in of poverty and gender concerns to AWM programs.

These messages need to be adapted to regional and local situations through a process of dialogue and study that will produce action programs. At the country level, the new World Bank Country Water Assistance Strategies can act as the locus for an integrated approach to AWM within broader sectoral and macroeconomic strategies.

POSTSCRIPT ON SUB-SAHARAN AFRICA

The case of Sub-Saharan Africa is raised in many parts of the report and deserves special attention: it is the poorest region, and growing poorer, yet with a large untapped endowment of water resources, Sub-Saharan Africa is where changes in AWM could make the biggest difference.

Hitherto, agricultural growth has been largely through extension of low-yielding, rainfed cultivation. The low infrastructure base, low capitalization, scant market development, and high levels of risk combine to keep farmers locked in a poverty trap of low-yielding, self-sufficiency strategies. Yet less than 5 percent of renewable water resources is abstracted and only

4 percent of agricultural land is under irrigation. Climate change and increasing hydrological variability increase the need for AWM—and improve its economic returns.

There are constraints—high cost, low population densities, weak skills base, and so forth—but it is clear that integrated investment in AWM infrastructure, markets, technology, institutions, and human development would help increase incomes and reduce poverty, offering Sub-Saharan Africa the prospect of the path to economic takeoff that Asian countries have so successfully pursued.

1

The Diversity, Contributions, and Achievements of Agricultural Water Management

Agricultural water management (AWM) is not a goal in itself but part of a process of resource management that provides a critical input to agricultural production and farmer incomes. Because of the way AWM affects development objectives across several sectors and is affected by them, the policy analysis and options explored in this report contribute to broader development of water resources, agriculture, socioeconomic welfare, and the environment, and to overall macroeconomic policy for growth.

This intermediary role results in strong links between AWM and the following key areas of public policy:

- *Water resource management.* In most developing countries, agriculture is the dominant user of water, accounting for more than 85 percent of all water use. Agricultural water use raises significant issues for water resource management—issues dealing with water scarcity, competing demands from other sectors, irrigation service delivery and system management, water use efficiencies, and so forth. The primary objective in coming years will be to balance water supply and demand among users to ensure adequate water for agriculture and sustainable irrigation system management while satisfying other needs.
- *Agriculture.* For many developing countries, agriculture is still the largest productive sector in the economy, the source of most economic growth and employment, and a large contributor to export revenues. Within the agricultural sector, irrigation is often the dominant contributor to value added, employment, and exports. Thus, agricultural policy issues affecting the irrigated sector—particularly those related to trade and incentives, input and output marketing and prices, investment, and food security—form an important part of overall agricultural policy. The key objective in the next decades will be to ensure that irrigated agriculture contributes to growth of sector value added and farmer incomes, and to food security at global, national, and household levels by meeting quickly rising demand for food at affordable prices.

- *Rural development.* Irrigated agriculture is an important driver of rural growth and an instrument of poverty reduction in the regions where it is developed. In addition, government rural development policies—rural infrastructure, rural incomes, socioeconomic development and poverty reduction, the role of women—have direct and indirect impacts on AWM policies and investments.
- *Environment.* AWM interacts in many ways, both positive and negative, with the environment. Environmental protection policy consequently exercises considerable influences on AWM, and vice versa. In coming years, reconciling environmental protection and sustainability with the need to intensify irrigated agricultural production will be a critically important goal.
- *Overall macroeconomic policy.* In most developing countries, the important contribution of the irrigated sector to national objectives of equitable growth through efficient resource allocation requires a close interaction between macroeconomic policy and AWM. Fiscal policy determines public investment, decentralization, subsidy, and cost recovery patterns. National policies on stakeholder engagement affect the shape of institutions and the distribution of roles among farmers, government, and the private sector. Income policies determine the nature of poverty reduction and social interventions. At the macroeconomic level, the tensions between policy objectives and political constraints set the overall political economy context for effecting reform in AWM.

1.1 THE DIVERSITY OF AGRICULTURAL WATER MANAGEMENT MUST BE RECOGNIZED IN ANY ANALYSIS OF THE SECTOR.

Of the world's total farmed area of 1.5 billion hectares (ha), about 18 percent (or 277 million ha in 2000) is irrigated. For developing countries as a whole, the irrigated area has almost doubled over the last 40 years to cover 234 million ha in 2000, representing about half the potential estimated by the Food and Agriculture Organization (FAO) (table 1.1). The South Asia and East Asia and Pacific regions account for 67 percent (or 157 million ha) of the irrigated area in developing countries. Within those regions, China and India account for 71 percent of the irrigated area. By contrast, very little irrigation development has occurred in Sub-Saharan Africa. Irrigated area in Sub-Saharan Africa is estimated at 7.1 million ha (AQUASTAT 2005), about equal to the irrigated area of Mexico and significantly lower than the irrigated area of Iran. Within Sub-Saharan Africa, Sudan, South Africa, and Madagascar alone account for 63 percent of the total irrigated area.

When analyzing the issues and assessing the performance of the sector, observers tend to consider AWM systems as being typically publicly funded and managed, large scale, irrigated from surface water sources, and pre-

Table 1.1. Irrigated Land Expansion by Region of the Developing World, 1961–2000

Region	1961–3 M ha	1979–81 M ha	2000 M ha	Annual growth rate (%)	Irrigated land as % of potential[a]
All developing countries[b]	**118**	**173**	**234**	**1.9**	**50**
Sub-Saharan Africa	4	5	7	2.0	14
Near East and North Africa[c]	13	18	21	1.7	62
South Asia	37	56	82	2.3	57
India	*25*	*37*	*58*	*2.6*	*65*
East Asia and Pacific	40	59	75	1.6	64
China	*30*	*45*	*55*	*1.4*	*70*
Latin America and the Caribbean	8	13	19	2.0	27
Europe and Central Asia	16	22	30	2.3	n.a.
World	**142**	**210**	**277**	**1.8**	**n.a.**

Source: FAO 2003d.

Notes: n.a. = not available.

a. FAOSTAT's estimates of irrigation potential area are based on individual country submissions of the area of land suitable for irrigation development, which, in turn, are based on available land and water resources and (often, but not always) on economic and environmental considerations. Wetlands and floodplains are usually, but not always, included.

b. "All developing countries" excludes Commonwealth of Independent States countries.

c. The Near East and North Africa, as defined by FAO, includes the World Bank Middle East and North Africa countries, plus Afghanistan, Turkey, and Cyprus.

dominantly planted to cereals or other relatively low-value field crops. This report describes a much more diverse sector. For example, in India and northern China, the area irrigated by groundwater rose from about 25 percent of the total irrigated area in the 1960s to over 50 percent in the 1990s. Groundwater irrigation exceeds 90 percent of the irrigated area in Saudi Arabia; 60 percent in countries as diverse as Bangladesh, Algeria, and the Republic of Yemen; and 50 percent in the Islamic Republic of Iran and Tunisia. In Brazil, irrigated areas are overwhelmingly private (90 percent), but almost entirely surface irrigated employing advanced technologies: 52 percent of the irrigated area uses modern irrigation methods such as mobile sprinkler systems, central pivots, microsprinkler, drip, and perforated tubes. In fact, AWM systems vary in many ways, depending on a large number of factors:

- *climatic conditions*, with significant differences in off-farm and on-farm water requirements and management, drainage needs, and cropping patterns according to whether the climate is humid, temperate, arid, and so forth;

- *water resources*, which can be surface water (from river or stream runoffs; lakes, lagoons, or dams; floods, or just rainwater) or shallow or deep groundwater (from hand-dug wells or tube wells) or both surface and ground (for conjunctive use);
- *size of the scheme,* which can be large-, medium-, or small-scale. The average size within each of these categories varies widely among countries: a 1,000-ha irrigation scheme is considered small-scale (or minor) in India, medium-scale in Morocco, and large-scale (or major) in Niger.
- *water conveyance and distribution method,* which can be through gravity (canal) or pressurized piped systems for conveyance, and through surface (canal, furrow, basin) or pressurized systems (sprinklers, drip, and so on) on farm;
- *management and institutional setup,* which can be public (government services or agencies), private (private individuals, companies, or water user associations), or a combination of both. Medium- to large-scale systems are usually built and run by governments, although increasingly with stakeholder participation, whereas small systems and groundwater systems are usually privately owned and operated.
- *irrigation water application method,* which can be full (satisfying the water requirements of the crops throughout their growing periods), supplementary (at critical development stages of the crops), or occasional (through rain water harvesting, diversion of intermittent runoff water or floods, and the like); and
- *type of crops*, which can be high-value cash crops such as fruits and vegetables, requiring high system reliability and flexibility, or less demanding, lower-value food crops such as cereals.

Table 1.2 presents the most common features that characterize publicly and privately managed irrigation schemes. It shows, for example, that in groundwater-irrigated areas, systems are generally smaller, predominantly privately funded and individually managed, and farming systems are more diversified with higher-value crops.

The nature of the AWM challenge surrounding water scarcity may also vary widely between countries and basins. In the Republic of Yemen and Mexico, for example, water scarcity is pressing and the primary challenge is to manage groundwater sustainably in the face of competing urban and agricultural water demands. In China and India, resource scarcity and very high agricultural use rates in many basins define the main challenges, but at the same time competing municipal and industrial demand is rising rapidly, and environmental issues of pollution and overdraft are threatening the resource base. In Pakistan, massive drainage problems and resultant salinization dominate the AWM agenda, while flood control is the major preoccupation in Bangladesh.

Table 1.2. Features of Publicly and Privately Managed Irrigation Systems

Feature	Publicly managed systems	Privately managed systems
Scheme size	Large scale	Small scale
Water sources	Surface	Groundwater
Water distribution	Collective	Individual
Water productivity	Lower	Higher
Drainage	Badly drained	Well drained
Cropping pattern	Less diversified	Highly diversified
Main crops	Lower value	Higher value

Source: Authors.

Within this wide variety of situations, analysis of AWM in its relationship to agriculture should focus primarily on those crops in each of the main climate zones that rely heavily on irrigation and the respective farming systems for those crops. The main crops that are mostly irrigated in the arid or sub-humid and the humid regions are presented in table 1.3.

1.2 IRRIGATED AGRICULTURE HAS BEEN VITAL TO MEETING FAST-RISING FOOD DEMAND.

This section discusses historical and current trends in demand and supply for agricultural products, particularly irrigated products. The section focuses on cereals, because cereals are the basic food commodity that the world must produce to survive and around which the struggle against hunger and poverty is coordinated.

Demand and supply for agricultural products have risen quickly.

World demand for food has more than doubled in the last 30 years, because of the expansion of world population. Developing country demand has almost tripled, outpacing population growth rates, as calorie intakes have increased. However, developing country per capita consumption of cereals is still only 40 percent of developed country consumption (247 kg per person annually in 1997–9, compared to 588 kg in developed countries) (figure 1.1). In some developing countries, particularly those in East Asia and Pacific, Latin America and the Caribbean, and the Middle East and North Africa, with lower population growth rates but higher growth in GDP, per capita consumption has gone up faster still.

Agricultural production in developing countries has risen enormously, almost keeping pace with demand. The pattern differs significantly by

Table 1.3. Irrigated and Rainfed Crops in the Developing Countries

Main crops	Arid and sub-humid regions[a]		Humid regions[b]	
	Mostly irrigated	Mostly rainfed	Mostly irrigated	Mostly rainfed
Cereals				
Rice	X	—	X	—
Wheat	X	—	—	X
Maize	X	—	—	X
Other cereals	—	X	—	X
Cotton	X	—	—	X
Sugar	X	—	X	—
Horticulture				
Fruits	X	—	—	X
Vegetables	X	—	X	—
Roots and Tubers				
Potatoes	X	—	X	—
Others	X	—	—	X
Pulses and oil crops	—	X	—	X

Source: Authors.

Notes:

a. Regions with annual precipitation of less than 1,000 mm.

b. Regions with annual precipitation over 1,000 mm.

region, however (table 1.4). The two most populous regions of the world—South Asia and East Asia and Pacific—have succeeded in improving production per capita and in maintaining high degrees of food self-sufficiency. This pattern also marks the two most populous countries in those regions, India and China. South Asia has balanced cereals supply and demand despite rapid growth of the population and rising per capita consumption (up from 162 kg annually to 182 kg annually from 1964–6 to 1997–9). The region achieved this remarkable outcome by a tripling of cereals production over the same 35-year period. These extraordinary results were achieved through investment in irrigation and widespread adoption of productivity-enhancing measures.

However, food imports are growing—and self-sufficiency ratios are declining—in several developing regions. Globally, the food self-sufficiency of the developing world has declined from about 95 percent in the mid-1960s to just above 90 percent at the end of the millennium. There has thus been a substantial shift in the location of production from the developing to the developed world, with accompanying foreign exchange and food security

Figure 1.1. Total Cereals Demand and Per Capita Consumption

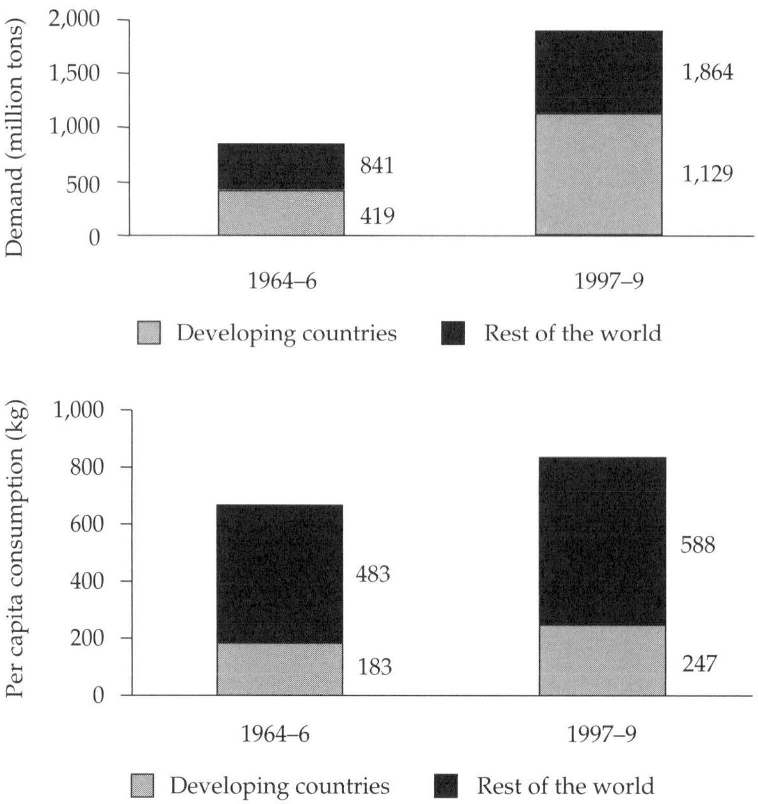

Source: FAO 2003d.

Table 1.4. Cereals Self-Sufficiency by Region (1997–9)

Region	Self-sufficiency %
Sub-Saharan Africa	82
Near East and North Africa	63
South Asia	102
East Asia and Pacific	95
Latin American and the Caribbean	88

Source: FAO 2003d, p. 68.

Figure 1.2. Per Capita Cereals Consumption by Region, 1997–9

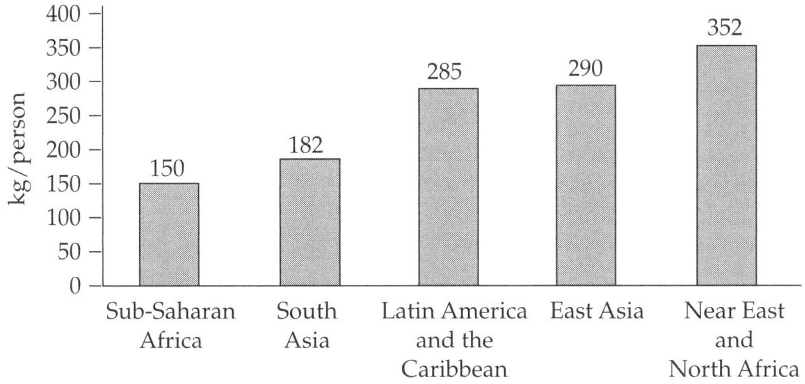

Source: FAO 2003d.

challenges for developing countries. The countries best placed to meet the challenges are those with fast-growing nonagricultural economies—particularly East Asia and Pacific—and those that have invested heavily in agricultural productivity, particularly in irrigation, as in South Asia, where 45 million new hectares of irrigated land have been developed since 1961.

Sub-Saharan Africa gives most cause for concern, with rates of increase in production considerably slower than population growth rates, and low levels of investment in irrigation and AWM. Of the 100 million hectares of new irrigated land developed between 1961 and 1999, only 2 million were in Sub-Saharan Africa. As a result, there is a growing cereals gap—food self-sufficiency has declined from 97 percent in 1964–6 to 82 percent in 1997–9. Yet, Sub-Saharan Africa has the least developed domestic food markets and is least able to afford imports, and is therefore increasingly vulnerable to shortages. In addition, Sub-Saharan Africa per capita cereals consumption is the lowest (figure 1.2), half that of East Asia and Pacific, reducing household food security further. The Near East and North Africa region has lower self-sufficiency ratios, but—unlike Sub-Saharan Africa—the countries of this region can afford to pursue the logic of comparative advantage and import an increasing share of their cereals.

Yield increases, especially of the irrigated crops, contributed the most to increased production.

Production of all major crops increased sharply over the last four decades. Cereals output in developing countries tripled over the 40-year period, to produce in 1997–9 an annual average of 1.2 billion tons on 440 million hectares. Production of other crops rose steeply, too. Irrigated crops, in particular, showed the largest increases. Crops that are predominantly irrigated—such as rice, wheat, maize, cotton, and vegetables—saw production increasing two- or threefold, and even fourfold in the case of wheat (figure 1.3). The production of irrigated fresh fruit and vegetables in developing countries increased fivefold from 1961–3 to 2002–4 (see figure 1.4). Yields have also increased dramatically, doubling in the case of vegetables. Developing country exports of fruits and vegetables have tripled in value over the last two decades, and fresh fruit and vegetables now account for over one-fifth of all developing country agricultural exports. The output of rainfed crops, by contrast, increased much more slowly, particularly the crops of marginal and semi-arid areas such as sorghum and millet.

More than two-thirds of the increase in crop production has come from yield increases, especially under irrigation conditions. FAO (2003d) estimates that more than two-thirds (71 percent) of the agricultural production increase in developing countries over the last 40 years came from yield increases, and 77 percent in total from "intensification" (that is, from increases in both yields and cropping intensity). Two main factors drove yields up much faster than irrigated area expansion: the widespread adoption of new varieties, inputs, and husbandry practices from the Green Revolution; and breakthroughs in irrigation technology such as the tube well that allowed easy groundwater extraction. While average yields for all cereals increased over this period by 80 percent (from 1.2 tons/ha to 2.8 tons/ha), yields of mainly irrigated cereals such as rice and maize more than doubled, and those of wheat went up threefold. Returns to water showed similar increases, with both rice and wheat more than doubling their yield per cubic meter (m^3). By contrast, the predominantly rainfed crops such as sorghum and millet recorded much lower increases in the

Figure 1.3. Production Indices for Mainly Irrigated and Mainly Rainfed Crops, 1997–9

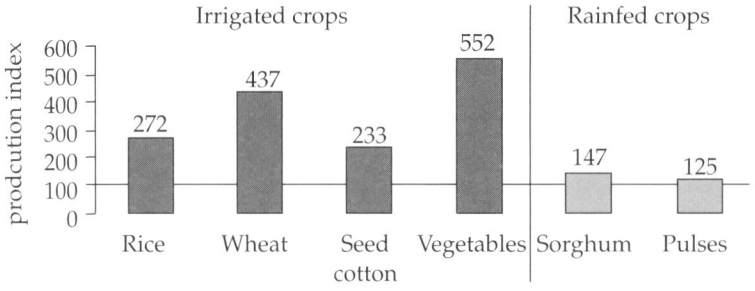

Source: FAO 2003d.
Note: 1961–3 = 100.

Figure 1.4. Increases in Production and Yields for Fruits and Vegetables in Developing Countries, 1961–3 to 2002–4

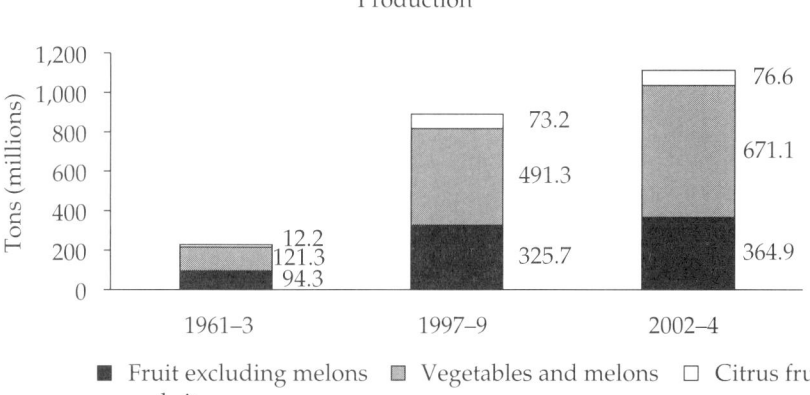

Production

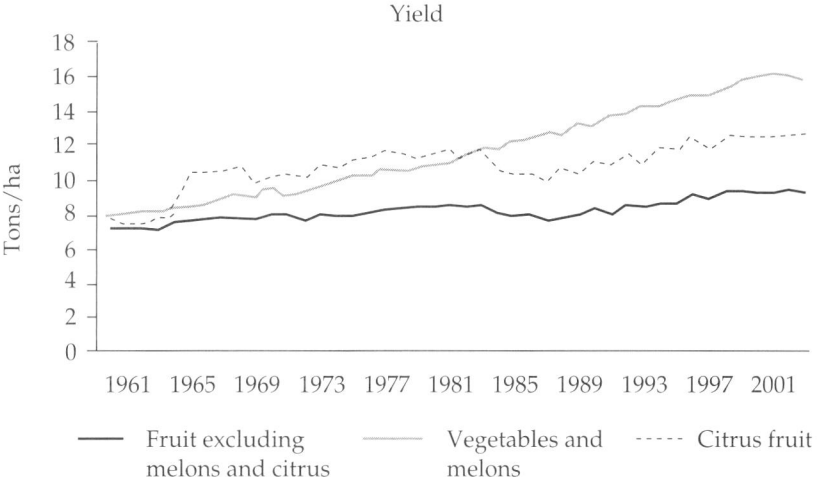

Source: FAO Agricultural Production Dataset (1961–2004).

yield index, of just 30 to 60 percent, reflecting the lower potential of rain-fed crops and the lack of any major research breakthrough so far. South Asia, where the percentage of arable area under irrigation is the highest, produced the most rapid growth in agricultural productivity. Some 80 percent of the region's rapid increases in production came from yield increases, and 94 percent from intensification overall, once increased cropping intensity is factored in (see figure 1.5).

Figure 1.5. Sources of Growth in Crop Production, 1961–99

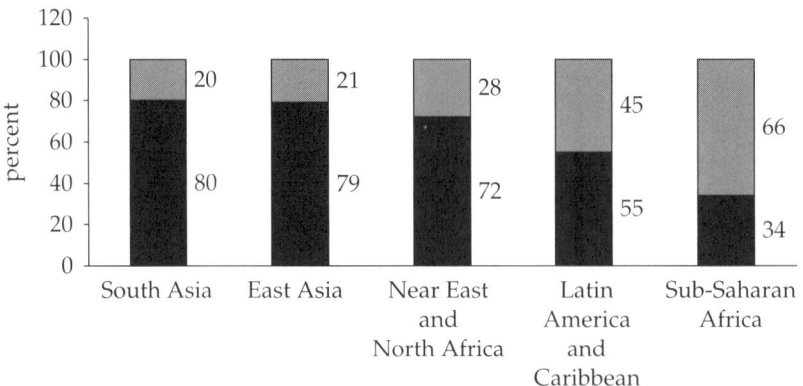

Source: FAO 2003d.

Only two regions—Sub-Saharan Africa and Latin America and the Caribbean—have expanded their arable area considerably. The exception among regions to the productivity revolution was again Sub-Saharan Africa, where low levels of investment in irrigation and AWM, combined with weak market development, left farmers in many countries with little option but to expand subsistence farming into new, marginal land areas. For example, Ethiopia, the second most populous country in Sub-Saharan Africa, has only about 170,000 ha of irrigated land developed, just 5 to 10 percent of its potential (see chapter 3). On existing irrigated lands in Sub-Saharan Africa, low levels of investment, weak markets, and the overriding subsistence farming objective of poor households have kept productivity low. Fertilizer use for irrigated rice in Madagascar, for example, averages 10 kg per hectare, one-thirtieth of world recommended levels, and paddy yields have stagnated at an average of 2 tons/ha over the last 40 years. Overall, yield increases in Sub-Saharan Africa have been low, contributing only one-third of production increases. In Latin America and the Caribbean, the rapid expansion of arable areas followed from the relative abundance of land resources and higher rainfall.

The increase in irrigated agriculture productivity contributed significantly to improving food security and reducing hunger.

The productivity of irrigated agriculture has increased significantly, especially when compared with rainfed agriculture. In 1997–9, irrigated production accounted for about 60 percent of cereals production in developing countries on 39 percent of the harvested land area (FAO 2003d). Water productivity (see below) has also increased considerably. Over the four decades, the amount of water needed to produce food for one person for a day halved from 6 m^3/day to less than 3 m^3/day. Over the same period, the production of rice and wheat went up by 100 percent and 160 percent, respectively, but with no increase in water use (FAO 2003d).

This massive productivity gain was the key element in improving food security and reducing world hunger. Between 1964 and 1999, average food intake in the developing world went up from 2,054 calories per day (cal/day) to 2,681 cal/day (figure 1.6). By 1997–9, average consumption in developing countries was just slightly below the world average of 2,803 cal/day. In East Asia and Pacific, (where China accounts for almost 70 percent of the population), per capita consumption has risen by half since the mid-1960s and now exceeds the world average. Several large developing countries that have developed their irrigation sectors intensively (including China, India, and Brazil) have brought average daily consumption for their people to 2,900–3,000 calories, a range at which the eradication of hunger is within the reach of each nation. The quality of diets has improved, too, with the absolute weight of cereals increasing in the consumption basket, but with more nutritious items taking an increasing share. Only in South Asia (with the exception of India, where irrigation is widely developed) and particularly in Sub-Saharan

Figure 1.6. Daily Per Capita Food Consumption

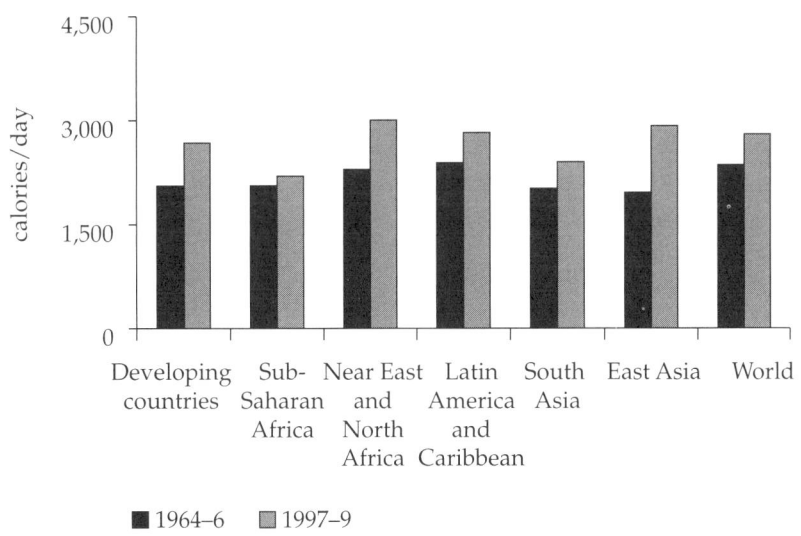

Source: FAO 2003d.

Figure 1.7. Incidence of Undernourishment in Developing Countries

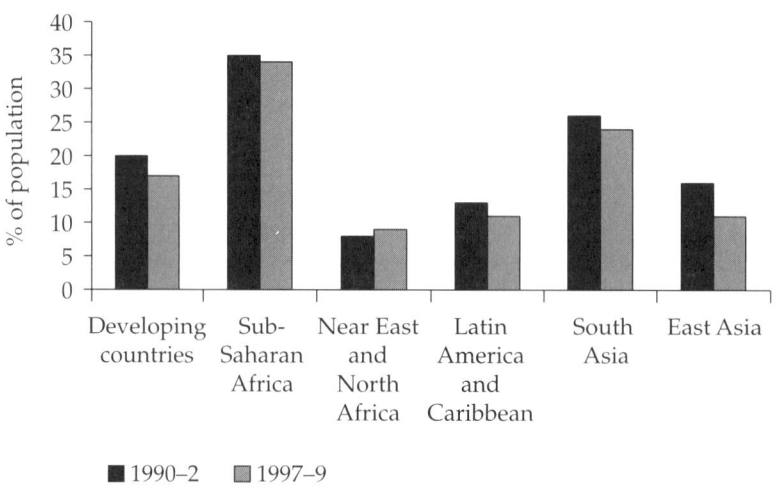

Source: FAO 2003d.

Africa are average consumption levels below world averages, with high rates of undernourishment and slow rates of decline in undernourishment (figure 1.7). In Sub-Saharan Africa, where irrigation development has been extremely limited, and where agriculture remains extensive and low yielding, more than one-third of the people are still undernourished.

1.3 AGRICULTURAL WATER MANAGEMENT HAS CONTRIBUTED SIGNIFICANTLY TO RURAL ECONOMIC GROWTH AND POVERTY REDUCTION.

Irrigated agriculture provides direct income, employment, and poverty reduction benefits, as well as many indirect benefits—not only economic but social, cultural, and environmental.

The benefits of irrigated agriculture extend beyond the primary crop production activity to affect the rural economy in which irrigation is developed.

In developing countries worldwide, irrigated agriculture can create significant employment in both farm and nonfarm activities. In Brazil, irrigation schemes covering 85,000 ha created over two jobs for every ha farmed, and a further four jobs in downstream activities and through the multi-plier effect from the incomes generated. In total, half a million new jobs were created (World Bank 2004a). In a study sponsored by the government of the Arab Republic of Egypt, a 5 percent agricultural growth rate (based on irrigation) created 1 million jobs—a quarter of a million in agriculture and three-quarters of a million in downstream and nonfarm production and services (Mellor and Gavian 1999). In Sub-Saharan Africa, the International Food Policy Research Institute estimates that every additional US$1 of new farm income increases total household income by between US$1.96 (in Niger) and US$2.88 (in Burkina Faso) (Delgado, Hopkins, and Kelly 1998). Many Asian countries have used irrigated agriculture as the mechanism to drive broadly based economic growth. In Malaysia's Muda irrigation scheme (100,000 ha), investment in improved irrigation and cul-tivation practices drove up farm incomes through double cropping and higher yields, providing farmers with more disposable income, which in turn stimulated rapid economic development in retail trade, service industries, and local manufactures. In the West Delta Irrigation Project in Egypt, the development of about 100,000 ha of irrigated lands is expected to lead to set-tlement of between 800,000 and 1 million people (including on-farm and off-farm labor, businessmen, service providers, and their families) (Hoevenaars and Slootweg 2004). When Mali's Office du Niger was revived in the 1990s, the local population rose by 100 percent as people were attracted by the myriad requirements of a growing economy (Aw and Diemer 2004).

Irrigated agriculture has also been critically important for poverty reduc-tion in many countries. In Asia, the poverty head count is typically far lower in irrigated areas—in Vietnam, 18 percent against over 60 percent in rain-fed areas. The Green Revolution achieved much of the reduction of poverty in Asia through the combination of water management with other factors to achieve good yields. The groundwater revolution also has a significant poverty reduction impact, bringing a reliable water source right onto the farms of poor people. A more detailed coverage of AWM and poverty reduc-tion is included in chapter 5.

Agricultural production is only one of the many services AWM provides to society.

In many parts of the world, AWM infrastructure has helped mitigate the impacts of droughts and floods, stabilize river flows, reduce silt loads, and so forth. Land leveling, water harvesting, and land and water conserva-tion practices have helped reduce soil erosion. AWM and its related hydraulic, marketing, and other rural infrastructure have helped to shape the countryside, increasing its amenity value and improving economic ser-vices to people. Through the age-old connections between people and water,

Box 1.1. Multifunctionality in Paddy Cultivation in Monsoon Asia

In addition to the economic and productive functions typically captured in a cost-benefit analysis, AWM provides numerous other services. In a study of paddy cultivation in monsoon Asia, these other functions were found to include

- *rural development functions*, such as the multiplier effects from higher disposable incomes;
- *environmental functions*, such as flood control (valued in Japan at US$16–24 billion), water filtration (valued in South Korea at US$1–5 billion), habitat values such as wetlands ecosystems, and ecotourism (for example, in Bali, where rural hotels are set in a waterscape of paddy fields);
- *social development functions*, such as the community solidarity and social capital built by the communal nature of much water management for paddy cultivation, and governance impacts from improved skill levels and experience in participatory management; and
- *cultural and religious functions*, such as religious rituals and cultural identity tied to the cycle of paddy cultivation. Other sociocultural values include landscape values; the cultural heritage of the constructed environment of hydraulic works, of farming practices, and of food and cooking; and the overall aesthetic value of an integrated environment and society.

Source: Adapted from Groenfeldt 2005.

AWM has strengthened social cohesion and enhanced cultural values and life in many regions. In addition, as demonstrated by several studies, AWM supports a wide range of other rural, social, cultural, and environmental development services, especially in the monsoon regions (see box 1.1).

One challenge for AWM is how to capture these multiple functions in accounting for costs and benefits. Conventional analysis gives preponderant weight to direct economic benefits (for example, resulting from incremental irrigated production) and ignores these externalities. Only in the developed world, particularly in Europe and Japan, have these values been formally incorporated into policy and practice. In both Europe and Japan, agriculture multifunctionality has been embraced as a means of expanding the interpretation of what constitute public goods that may need public financing.

2

The Challenges Facing Agricultural Water Management

This chapter discusses the main problems affecting agricultural water management globally, including the problems of water availability, the challenges of irrigation management, the challenge of rainfed agriculture, the scope for technological change, and the interaction of agricultural water management with the environment.

2.1 THE RATE OF IRRIGATION EXPANSION IS SLOWING DOWN.

The global irrigated area grew from 142 to 276 million hectares (ha) between 1961–3 and 2000, doubling in 40 years. However, the pace of development was faster in the early years and slowed considerably in the later years: growth rates of around 2 percent a year in the 1960s and 1970s slowed to 1.5 percent in the 1980s and to hardly 1 percent in the 1990s. Worldwide, per capita irrigated area peaked in 1978 (0.045 ha per person), but has since fallen 5 percent, and continues to decline as population rises and as few new irrigated areas are developed (Postel 1999). This slowdown is reflected in the decreasing rate of dam construction. From the 1950s to the mid-1970s, about 1,000 new, large dams were constructed each year. By the early 1990s, only 260 dams, on average, were being built each year (Postel 1999). For decades, however, this slowdown in surface irrigation development was compensated for by the rapid growth of groundwater irrigation.

Although there are no reliable statistics on global irrigation financing, the best estimate is in the range of US$30 billion to US$35 billion a year, for both investment and operations and maintenance (see table 2.1). This level of investment is considerable—equal to that in drinking water supply and sanitation and hygiene, for example—but evidence from government and international agency financing suggests that investments for irrigation, drainage, and broader agricultural water management have been following a downward trend. Lending for irrigation and drainage by multilateral development institutions declined from a peak of about US$3.0 billion annually in the mid-1980s to about US$2.0 billion in the mid-1990s. (Cleaver and Gonzalez 2003; Winpenny 2005). World Bank lending also declined.

Irrigation and drainage accounted for 7 percent of World Bank lending for the 30-year period 1960–90, higher than any other single sector. From 1950 to 1993, the Bank lent roughly US$31 billion (in 1991 dollars) for various forms of irrigation and drainage in 614 projects. Bank lending for irrigation and drainage in the 1970s and 1980s accounted for 10 percent of total lending, with an annual average of over US$1.1 billion (1991 prices). Since then, lending in the sector has dropped considerably, reaching a record low of US$220 million in fiscal year 2003. Lending increased, however, to US$769 million in FY2004 and to US$1,069 million in FY2005.

This decline in investment reflects a broader decline in public spending for agriculture as a whole (World Bank 2005a), but also reflects factors specific to irrigation that have changed both the scale and the nature of investment. As the World Water Council commented, the era of new costly large scale public capital irrigation investment is almost at an end—now more efficient, user managed rehabilitation, and operation and maintenance are likely (World Water Council 2003).

Certain physical and technical factors have played a role. In many countries, the potential for expansion is limited: the water resources of numerous rivers have been fully exploited, with some basins, such as the Hai basin in China, already withdrawing over 100 percent of the renewable resource (including mined groundwater), with no water reaching the sea.[1] On the economic and financial side, in addition to the lackluster trend of agricultural commodity prices, the unit cost of development has increased, and the resulting perception of low economic returns has deterred investment.[2] Perceptions of poor performance of irrigation investments have been heightened by often overoptimistic assumptions during planning and design.

Studies over the last 20 years show a rise in the cost of irrigation development due to increases in the costs of civil works and equipment and to the increasing difficulty of new sites. The most recent study (IWMI 2005), which compares global data from 314 irrigation projects, shows an aver-

Table 2.1. Indicative Annual Investment in Water Services for Developing Countries

Water use	US$ billion
Drinking water	13.0
Sanitation and hygiene	1.0
Municipal wastewater treatment	14.0
Industrial effluent	7.0
Irrigated agriculture	32.5
Environmental protection	7.5
Total	**75.0**

Source: World Water Council 2003.

age cost of US$6,600 per ha in 2000 prices for new construction and US$2,900 for rehabilitation. The study confirms that costs are highest in Sub-Saharan Africa, at US$14,500 for new construction and US$8,200 for rehabilitation.

Perhaps the single most important factor is simply that investment policies have changed, both within irrigation and between sectors. Recent investment patterns within irrigation have aimed more toward increased efficiency and sustainability of water use, for example, in system rehabilitation, improvement, and operation and maintenance; and in management, institutions, and policy, all of which have lower investment costs than the capital-intensive works of system development. Between sectors, governments and agencies have focused more on general budget support for and investments in social development (education and health, for example), privatization, and the environment than on irrigation development.

2.2 WATER AVAILABILITY FOR IRRIGATED AGRICULTURE IS INCREASINGLY CONSTRAINED.

Population growth and nutritional improvements are driving up demand for agricultural water. World population growth has begun to ease in recent years, but developing country populations overall are still increasing at an annual average of 1.4 percent (2.6 percent in Sub-Saharan Africa). Some 80 million people are added to world population each year, almost all in developing countries. This quickly growing population is eating better than ever before and driving up demand for agricultural products and for the water that produces them. For example, as income levels rise, dietary demand for meat rises (that is, shifting from Diets 3–6 to Diets 0–2 in table 2.2). The water required to produce one ton of beef is five times that required to produce one ton of soybeans (table 2.3). The growth in meat consumption, particularly marked in Asia, especially in China, thus increases the water requirement for food production.

Table 2.2. Virtual Water Content of Diets

Diet	Water content (m^3/person/day)
Diet 0 (reference United States)	5.4
Diet 1 (5% reduction in animal product)	4.6
Diet 2 (poultry replaces 50% of beef)	4.8
Diet 3 (vegetable products replace 50% of red meat)	4.4
Diet 4 (50% reduction in animal products)	3.4
Diet 5 (vegetarian)	2.6
Diet 6 (survival)	1.0

Source: Zimmer and Renault 2003. Reprinted with permission from Elsevier.

Table 2.3. Virtual Water Content for Selected Products

Product	Water content(m^3/ton)
Beef	13,500
Pork	4,600
Poultry	4,100
Soybeans	2,750
Eggs	2,700
Rice	1,400
Wheat	1,160
Milk	790

Source: Renault and Walender 2000.

Water availability for irrigated agriculture is decreasing. Irrigated agriculture accounts for three-quarters of world water withdrawals from surface and ground water and 85 percent of consumption of water withdrawn in developing countries.[3] This large share does not in itself mean scarcity, because in many countries withdrawals take up only a relatively small share of potentially available resources. On average, developing countries withdraw only 7 percent of their total renewable water resources for irrigation (table 2.4). However, withdrawal percentages differ widely by region, from 1 percent in Latin America and the Caribbean to 53 percent in the Near East and North Africa, which faces the most pressing scarcity problem. With the rapid development of irrigated agriculture over the last four decades,

Table 2.4. Renewable Water Resources and Irrigation Water Requirements in Developing Countries

Annual averages 1997–9

	Sub-Saharan Africa	Near East and North Africa	South Asia	East Asia	Latin America and the Caribbean	All developing countries
Precipitation (mm)	880	181	1,093	1,252	1,534	1,043
Renewable water resources (km^3)	3,450	541	2,469	8,609	13,409	28,477
Irrigation water withdrawal (km^3)	80	287	895	684	182	2,128
Irrigation water withdrawal as percentage of renewable water resources	2	53	36	8	1	7

Source: FAO 2003d.

water withdrawals for irrigation have increased by more than 200 percent. At the same time, demand from other sectors has increased even faster, especially for municipal and industrial uses, as populations increase and urbanization gathers pace. As a result, per capita water availability continues to decline in developing countries (figure 2.1).

Many countries and basins are already "water scarce"—the threshold for scarcity is considered to be withdrawals of 40 percent of renewable water resources. Experience has shown that beyond that point, costs rise sharply, groundwater is depleted, and conflict between agriculture and competing municipal and industrial water uses intensifies. In two regions (South Asia and the Near East and North Africa), withdrawal rates are already in excess of 30–40 percent and countries in those regions are experiencing scarcity. Worldwide, at least 10 countries already use more than 40 percent of their resources in irrigation, and an additional 8 countries use more than 20 percent. In some countries and basins, scarcity is acute. When all uses are taken into account, the Arab Republic of Egypt is using 99 percent of its renewable resources, and in the Hai basin in China, use has reached 140 percent of renewable resources (Rosegrant, Cai, and Cline 2002b).

Figure 2.1. The Decline of Water Availability in Developing Countries

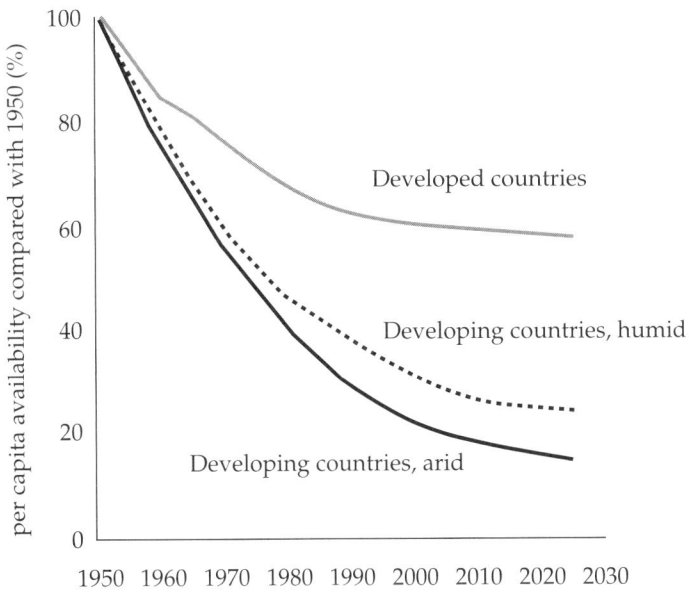

Source: World Bank 2002a.

2.3 THE GROUNDWATER IRRIGATION REVOLUTION HAS BEEN ACHIEVED AT THE PRICE OF THE DEPLETION OF THESE RESOURCES IN MANY REGIONS.

The rapid growth of groundwater use in recent years brought massive benefits to farmers, but also resulted in widespread overexploitation of groundwater resources. In some basins, water tables have been falling at an alarming rate. In large areas of India and China, groundwater levels are falling by one to three meters a year.

Overdraft and the accompanying deterioration of groundwater quality have been driven in many countries by lack of an institutional framework that can moderate use. At the same time, many countries have created an incentive structure that encourages overdraft, particularly through subsidized energy prices (India, the Republic of Yemen). In many areas, rapid depletion is causing profound social and environmental impacts because both water supplies and agricultural employment are threatened. Notable examples of depletion and water quality decline include the coastal aquifers of Gaza, Gujarat (in India), west Java, and Mexico.

Groundwater depletion can lead to irreversible land subsidence, saltwater intrusion, and pollution of the water resource in addition to increased pumping costs. Ultimately, if depletion continues unchecked, the groundwater resource will become exhausted or too deep to extract economically. Regions affected include some of the world's major grain-producing areas such as the Punjab and the North China plain. Globally, the groundwater problem presents an important risk to world food production—now about 10 percent of the world's food is produced using mined groundwater. The poor are particularly vulnerable, because the richer farmers can pump out deeper and faster. (More detailed coverage of the groundwater issues is included in chapter 5.)

2.4 PUBLICLY MANAGED IRRIGATION SCHEMES HAVE GENERALLY PERFORMED POORLY.

Performance of irrigated agriculture in publicly managed schemes (which cover about half the irrigated areas in the developing countries) generally falls well below technical and economic potential. The performance of the large-scale irrigation (LSI) schemes has been particularly disappointing. In most of these schemes, farmers often receive poor water service, and reliable and timely irrigation service delivery is the exception rather than the rule. The likely causes of these disappointments are many. The decline in agricultural product prices and the limited access to high-value product markets have restrained both diversification and investment in water use efficiency in many schemes.

However, the major causes of poor service delivery are commonly located in the interrelated problems of bureaucratic institutional setup and rigid technical design, both of which generally originate in the top-down, planning-led approach to irrigation. Bureaucratic institutional setups for LSI have contributed to poor service delivery in a vicious circle of insufficient funding, inadequate operation and maintenance, and system deterioration, often leading to the need for successive rehabilitations. Technical design has suffered from the same top-down approach. Many schemes were constructed with inflexible delivery patterns, which are suitable to deliver water according to preset schedules, but are incapable of responding to changes in demand by the users. Designers typically have been unfamiliar with the constraints of farmers and have paid insufficient attention to how schemes can be operated. As a result, most publicly managed schemes achieved neither fiscal efficiency nor demand-responsive water service. Even in the United States, the Bureau of Reclamation now recognizes that many of the irrigation schemes developed in the past cannot be operated efficiently.

This top-down approach arose because governments in the past shouldered the burden of investment in the large-scale irrigation sector, and subsidized both infrastructure and inputs, especially water and power (for pumping). Public policy has generally taken the physical endowment of land and water as a "national potential" that needs to be realized at almost any price. Thus, supply-led approaches predicated upon LSI infrastructure dominated, leading to neglect of market signals and a sharp discontinuity in policy, institutional capacity, and investment for the provision of adequate irrigation management and ancillary agricultural services.

The results have been mixed at best, with some highly efficient, sustainable systems, and others plagued by poor water service and deterioration of assets. Some large-scale systems were poorly designed, with insufficient provision for drainage and consequent soil degradation. The resulting schemes have often been unable to cover costs, creating a heavy fiscal burden for government and leading to disappointing economic returns on investment.

Central bureaucracies and public sector irrigation institutions have often lacked the structure and incentives to optimize productivity.

In most developing countries, LSI schemes have been managed by state bureaucracies and rigid, formal irrigation institutions. Under such structures, system management often fails to respond to the needs of users, particularly of smallholders. Cost recovery is low and water-use fees are not collected fully and are not allocated to proper system operation and maintenance. Irrigation institutions are often not adequately equipped to adapt to changing circumstances and expectations, and suffer from bureaucratic incentives and from institutional rigidity. All this results in poor water delivery, high fiscal costs (up to the mid-1990s, irrigation and drainage absorbed more than half of all agricultural investment in Pakistan, China, and Indonesia), deteriorating systems, and poor economic and production performance. (See box 2.1 for the World Bank's response to problems with LSI.)

Large-scale schemes have not adjusted well to changing market signals.

The production of low-value cereal crops (particularly wheat, corn, and sorghum) for which many LSI systems were constructed is becoming less competitive with rainfed production in the global marketplace. Irrigation systems are expensive to construct, operate, and maintain, and irrigation requires considerable labor input. At the same time, international prices of these low-value commodities have faced continued pressure. Mexico, for example, is presently struggling with these issues—more than half of its irrigated area of 6 million ha is dedicated to producing wheat, corn, and sorghum, crops that will be hard pressed to compete with rainfed production from northern neighbors when trade in these commodities is completely liberalized in 2008. The North American Free Trade Agreement and proximity to the U.S. market open the possibility for irrigated farmers to produce horticultural products and so improve incomes. However, in many cases this diversification will require wholesale modernization of irrigation systems.

Box 2.1. Spotlight on Large-Scale Irrigation Management

The Bank's Water Resources Sector Strategy (WRSS) considers the modernization of irrigation agencies to be its top priority for the sector, with an agenda that includes

- separating bulk infrastructure from distribution infrastructure;
- separating the public from the private aspects of the system;
- clarifying the public and private roles for service delivery; and
- devising a set of responsibilities and incentives linked to the key output—quality water service.

Source: Authors.

2.5 WATER MANAGEMENT FOR RAINFED AGRICULTURE HAS BEEN NEGLECTED.

Rainfed farming is a priority—not only does rainfed agriculture account for 60 percent of current agricultural output in developing countries, but rainfed areas are home to most of the world's poor. Rainfed farmers typically farm marginal lands or dry lands, with little or no access to a controlled water source. The Green Revolution largely bypassed these farmers. Improvements in yields and water management have been scant, and the growth of rainfed production through extension of the cultivated area has kept incomes low and has harmed the environment. Particularly in Sub-Saharan Africa, the availability of land for extensification has contributed to the neglect of options to move up from rainfed agriculture to access to controlled water sources—from simple supplementary irrigation to the development of large-scale schemes. As a result, rainfed farmers in Africa have remained very poor. Sub-Saharan Africa is the only region where poverty has increased in recent years. The technological options for improved water management in rainfed agriculture have been piloted and developed over the last decade, but are seldom implemented on a large geographical scale. Detailed discussions of water management for rainfed agriculture are included in chapters 4 and 6.

2.6 TECHNOLOGY IS AVAILABLE, BUT OFTEN IS NOT DISSEMINATED AND ADOPTED.

One finding of *Shaping the Future of Water for Agriculture: A Sourcebook for Investment in Agricultural Water Management* is that "more technology is available than we know what to do with" (World Bank 2005b, p. 10). Many innovations are possible in water management, agriculture, and ecological management that would improve productivity or conserve water. Drip technology, for example—which since the 1970s has shown its ability to produce high yields per unit cost of water—has been adopted on less than 1 percent of irrigated lands worldwide, although investment costs have become affordable. Experience is that farmers may well be aware of technological options, but do not invest in technology unless pushed by cost incentives (rising water prices, for instance) or pulled by profitable market opportunities. Plainly, both technology transfer and market development and incentive questions need to be addressed. In addition, farmers have to minimize risk, particularly risk of access to adequate quantities of water at the right time. Thus, secure water entitlements and demand-responsive water service are key factors in encouraging farmers to invest in new technology. Beyond that, continuing research is needed to improve returns to water and increase farmer incomes. The agenda includes research on crops tolerant of salt and drought, on basic food grains, and on improving water management in rainfed areas.

Economic and social aspects are often subordinate to production aspects. In most countries, irrigation development has been driven by food supply imperatives—particularly food self-sufficiency considerations—rather than by market demand. As a consequence, LSI schemes have generally been built to produce low-value staples. Not surprisingly, users—especially small ones—cannot pay the full financial cost of water, let alone the opportunity cost. Despite the strong arguments for recovery of the cost of supplying irrigation water (incentives to water productivity, covering scheme costs, environmental concerns, equity), in reality, irrigation is heavily subsidized almost everywhere. Substantial distortions exist on the revenue side, too: commodity prices are kept low by patterns of producer support in the Organisation for Economic Co-operation and Development countries and by patterns of consumer and producer support in the developing countries. The tendency is to build complex webs of countervailing subsidies that constrain market development and further distort incentives to efficiency. Meanwhile, producers who have no alternative to producing staples to make a decent living face difficulties, especially in countries where crop yields are generally low (as in Sub-Saharan Africa). On the social side, despite the contribution of agricultural water management to poverty reduction (see chapters 1 and 5), few investment programs explicitly target the needs of the poorest by taking into account key factors affecting their livelihood, such as distribution of land holdings, security of water entitlements, the vulnerability of irrigation tail-enders (that is, irrigators whose plots lie at the bottom end of the water distribution system and who receive only residual water), and the appropriateness of technology to the situation of poor households. In addition, as described in detail in chapter 5, very few programs have taken gender aspects into account.

2.7 THE ENVIRONMENTAL IMPACTS OF AGRICULTURAL WATER MANAGEMENT HAVE BEEN NEGLECTED.

The multiple beneficial impacts of agricultural water management on the environment and on society were discussed earlier in this volume. Rural people are the trustees of much of the world's land and water resources, and thus are central to achieving the sixth Millennium Development Goal, to "ensure environmental sustainability" (see World Bank 2005c). However, this trusteeship is increasingly hard to respect as countries approach the limits of water and land resources. The resulting stresses create environmental risks. Agriculture is by far the largest user of land and water resources. Over 50 percent of the surface area of the major river basins in South Asia is covered by agricultural activity, as is over 30 percent of the major basins in Latin America and the Caribbean, North Africa, and East Asia. In many developing countries, irrigation water withdrawals exceed 90 percent of the total. Inevitably, the tension between agricultural production and protection of natural resources has grown. Environmental costs and risks of irrigated agriculture have become clearer: land degradation, salinization, and erosion; loss of environmental water flows; pollution; destruction of natural habitats and livelihoods through drainage of wetlands and through land expansion and deforestation; and waterborne disease.

Much of the world's irrigated land suffers from drainage problems.

During the rapid expansion of irrigation over the last 40 years, drainage was largely neglected. As a result, irrigated land in many large-scale schemes has become waterlogged and salinized due to the rise of the water tables and accumulation of salts. Waterlogging and salinization have become constraints to productivity. In India, for example, waterlogging affects 8.5 million ha and results in the loss of 2 million tons of grain each year. These problems are largely due to poor design and management of irrigation systems. Few schemes consider drainage needs in their design, while preset irrigation schedules—and sometimes uncertainty over future deliveries— encourage farmers to over irrigate. Often the pricing structure does not encourage water saving.

Worldwide, half of all existing soils are affected by salt to some degree. Build-up of salts through irrigation affects as much as 20 percent of the total irrigated area. In some semi-arid countries, up to half of the irrigated area is affected, with average yield decreases of 10–25 percent. The problem can have a large impact on countries and on local economies. Some major food-producing areas of the world are seriously affected, including the western Punjab and the Indus Valley. India has already lost 7 million ha of productive land to salinization. More detailed coverage of drainage is included in chapter 6.

Investment in drainage is low, despite good returns.

Worldwide, at least 20–30 million ha of irrigated land require drainage investments, and the need is growing at up to one-half million ha each year. One-half million ha go out of production each year. Drainage is a good investment: projects have generally produced good rates of return and improved farmer incomes. The cost of "saving" an irrigated ha through drainage is generally less than US$1,000, compared with more than US$6,000 to create a new irrigated ha. Yet, investment has dwindled as projects have focused on upstream

Table 2.5. Global Distribution of Cropland and of the Percentage of Land Drained

Region	Total cropland (1,000 ha)	% of land drained
Africa	49.4	1.3
Asia	491.3	9.4
Eastern Europe	170.2	12.5
Latin America	150.3	5.5
Middle East and North Africa	69.4	12.1

Source: World Bank 2004b.

irrigation and farming. In most developing countries, less than 10 percent of land is properly drained (see table 2.5). The current rate of subsurface drainage development is only 100,000–200,000 ha per year.

Overall land degradation has reduced productivity. Land degradation caused by agricultural water management practices and by lack of drainage is affecting some of the world's most fertile basins (see table 2.6.) The global cumulative loss of cropland productivity from all sources of degradation since 1945 has been estimated at 13 percent but in Sub-Saharan Africa, estimates are 25 percent and in Latin America and the Caribbean, as high as 38 percent. A recent study (Byerlee, Xinshen Diao, and Jackson 2005) questions whether technological gains in the Pakistani Punjab can be sustained because of the severe degradation of land and water resources. In addition to loss of on-site productivity, the resulting off-site impacts are severe, usually more costly than the on-farm impacts—siltation of streams and reservoirs, loss of fish productivity, rising water storage costs, and incidence of flood damage.

Table 2.6. Major Production Basins Affected by Land Degradation Due to Salinity

Region	Production Basin
South Asia	Indus River basin
Middle East and North Africa	Tigris and Euphrates River basins, Jordan River basin, Nile Delta
East Asia and Pacific	Yellow River basin (Ningxia and Shandong provinces of China), Mekong River basin (Thailand and Vietnam)
Sub-Saharan Africa	Limpopo River basin , Volta River basin
Latin America and the Caribbean	Northern Mexico, Andean Highlands
Europe and Central Asia	Amu Darya and Sir Darya basins
North America	Colorado River basin

Source: Abdel-Dayem 2005.

Water withdrawals for irrigation have ignored environmental and health impacts.

Impoundments such as dams and abstractions for agriculture and other sectors have profoundly modified the flows of most of the world's rivers. These interventions have had significant impacts, reducing the total flow of many rivers and affecting both the seasonality of flows and the size and frequency of floods. In many cases, these modifications have adversely affected the ecological and hydrological services provided by water ecosystems. In some basins, water no longer reaches the sea, and environmental flows have virtually ceased. Excess withdrawal has lowered water quality, finally reducing water supply for human uses. These modifications have increased the vulnerability of people—especially the poor.

Irrigated agriculture is a source of pollution in many regions. The use of chemical inputs has been a central part of the productivity revolution, particularly in irrigated agriculture, but has also resulted in growing levels of pollution. Irrigated agriculture is the main source of nitrate pollution of groundwater and surface water, as well as the principal source of ammonia pollution. Fertilizer and pesticide uses are polluting both water and the atmosphere: nitrogen and phosphate enrichment has led to eutrophication (a process whereby water bodies receive excess nutrients that stimulate

---

Box 2.2. Irrigation Water Quality and Health in Egypt

Egypt's Ministry of Water Resources and Irrigation estimated the economic costs of impaired water quality at up to 65 billion Egyptian pounds (US$11.2 billion) annually. Health impacts far outweighed all other costs from poor water quality. These included an increase of 5–40 percent for various forms of cancer and heart disease in areas that irrigate with drainage water. The largest problems came, however, from nonagricultural pollution, particularly municipal and industrial waste discharges into waterways, and from lack of adequate sewerage in rural areas.

Solutions identified include improved municipal sewerage and wastewater treatment, cost recovery for urban sanitary services, and pretreatment of waste by industries. At the local level, the recently established water boards are taking action on canal cleaning and the creation of safe washing places. Most important in these initiatives is the participation of stakeholders through information disclosure and local action plans, for example, on domestic sanitation in rural areas.

Source: Hoevenaars and Slootweg 2004.

excessive plant growth). Both the environment and human health are detrimentally affected (box 2.2). Worldwide, pesticides contribute to an estimated 26 million human poisonings and 220,000 deaths each year. The problem is likely to grow in the developing world. Several large developing countries have fertilizer and pesticide application rates already exceeding those that caused serious environmental damage in developed countries.

Uncontrolled agricultural water use may affect health. Waterborne infections account for 90 percent of all human infectious diseases in the developing world. Disease related to irrigation and agricultural water management is a part of the problem. Schistosomiasis, caught from snails present in canals and drains, is contracted by more than 200 million people annually and causes an estimated 20,000 deaths. Mosquito-borne malaria infects more than 2.4 billion people and kills 2.7 million people each year.

Wetland reclamation for irrigation development has affected natural habitats and livelihoods. Clearance of wetlands has damaged hydrological functions such as groundwater recharge and, by changing natural habitat, has reduced biodiversity. By reducing floodwater storage capacity, drainage of wetlands has also contributed to flood damage. In central China, around a half million ha of wetlands have been reclaimed for crop production since 1950. While this has boosted irrigated production, the flood storage capacity of the wetland system has shrunk to one-third its original capacity. This contributed to the flood disasters of 1998, which caused damage estimated at US$20 billion (FAO 2003d).

3

The Changing Global and National Contexts for Agricultural Water Management

Agricultural water management (AWM) must contribute to the production of the greater quantities of food and fiber required to feed and clothe growing populations. From the discussion of constraints in chapter 2, it is clear that in most countries, much of this growth cannot come from mobilizing additional land and water resources, but must come from getting more out of less—more crop, cash, and jobs per drop. This chapter discusses changes in the global and local environments that affect these challenges for AWM.

The chapter highlights how global debate on water resources management, on food security, and on trade is sharpening the agenda for AWM. Research for AWM is embarking on changes to match this agenda. A practical dialogue between the international community and developing nations is prompting changes in the governance of the irrigation sector, in investment patterns, in the way that investment is financed, and in the treatment of the environment. At the national level, some governments are introducing changes in policies and practices concerning the roles of respective stakeholders. Drivers of change are often internal, the product of evolving paradigms of the way in which governments and other stakeholders interact in the development process, but external drivers, such as resource scarcity and degradation or climate change, also play an important role.

3.1 CHANGES IN THE GLOBAL DEVELOPMENT CONTEXT ARE AFFECTING AGRICULTURAL WATER MANAGEMENT

This section examines how changes in the global development context affect AWM—how water resources management has become a global issue, and how, after relatively little attention, AWM and water for food are only now starting to receive the international attention their strategic role in the world's future merits. More broadly, the section examines the interactions between AWM and the global debates on food security and hunger alleviation, and on trade in agricultural products. The section also looks at the international research agenda on AWM.

Water resources management is increasingly a global issue.

Water resources management has become the subject of intense international debate. The Dublin Conference of 1992 set basic principles that have guided thinking and practice in water resources management. The Dublin *institutional principle* established participation of all stakeholders (from governments to women and the poor) and decentralization as the best practice for water resources governance. The *instrument principle* highlighted the policy implication of growing water scarcity—the need for demand management through an incentive structure that reflects the true value of water to society. Finally, the *ecological principle* established the goal of integrated, intersectoral management of the resource and the need to factor environmental considerations into water resources management. Subsequent conferences—especially the meetings of the World Water Forum in Marrakesh in 1998, The Hague in 2000, and Kyoto in 2003—confirmed the basic principles set out at Dublin in light of worldwide implementation experience. The principles and their application to water resources management globally and nationally, including AWM, are supported by two international policy partnerships: the World Water Council, "the world's water policy think tank," which publishes the influential journal *Water Policy*; and the Global Water Partnership, which translates recommendations for action on water management into specific services for developing countries and which developed the Framework for Action, a set of strategies, mechanisms for implementation and priorities for short-term action and investment.

In general, these conferences and their supporting partnerships have provided space where experience, research, and principles can interact and evolve, consolidating best practice and confirming directions. The Dublin principles do, in fact, underlie many important changes happening in the AWM sector, including

- the growth of decentralized and inclusive governance models in the irrigation business—including participatory irrigation management and irrigation management transfer;
- awareness of the role of water as an economic good and of the need to move toward cost recovery and to use economic mechanisms to transfer water to the uses that society most values;
- the call to integrate irrigation in basinwide approaches; and
- recognition of the trade-offs between AWM and the environment.

The policy-based forums are supported by several international capacity-building partnerships. The International Commission on Irrigation and Drainage (ICID) is a scientific and technical nongovernmental organization (NGO), dedicated to improving water and land management through training, research, and development. The International Program for Research in Irrigation and Drainage (IPTRID) is an independent, multidonor trust fund program created by the World Bank in 1992 to support irrigation technology research, transfer, and adoption. Following a move to the Food and Agriculture Organization (FAO) in 2000, IPTRID now provides capacity-building and advisory services to help developing countries formulate irrigation and AWM programs within poverty-reduction strategies. These activities overlap with those of the International Water Management Institute (IWMI) (described below), and there is a case for renewing IPTRID's former mandate in researching institutional and technical ways to improve the management of large-scale irrigation. The International Network for Participatory Irrigation (INPIM) was launched by the World Bank in 1995. It has since become an independent nonprofit organization whose mission is to help analyze and disseminate experiences, support pilots, and build capacity and country ownership for irrigation-sector institutional reforms, with particular focus on the stakeholder participation agenda.

Food security and hunger alleviation are increasingly global issues.

Global agreements on food and hunger have focused world attention on the food production challenge (World Bank 2005a). Beginning with the 1996 World Food Summit and the resulting Rome declaration and Plan of Action, the international community has kept the problem of food security and the agenda for hunger alleviation at the forefront of debate. The Millennium Development Goals have underlined the central challenge of increasing food production and reducing hunger and malnutrition. The practical implications have been under study by the UN Hunger Task Force, set up by the Secretary General to examine in more detail the steps needed to lower the incidence of hunger in an economic and sustainable way. The task force reported in late 2004, finding that the Millennium Development Goal of "halving hunger" by 2015 is attainable, and is an important milestone in the global effort to eliminate hunger completely. The task force has, among other things, highlighted the key role of water, emphasizing that many of the world's most hungry people are found in the arid and semi-arid tropics, where water availability is critical. The task force report discusses the role of water harvesting and small-scale irrigation combined with efficient water use in transforming crop and livestock production, and recommends developing these techniques and promoting their adoption on a broad scale. The report argues that better small-scale water management can make virtually every other operation on a farm more productive and less risky, and cites communities in India where a combination of water harvesting and the rehabilitation of degraded land has boosted farmers' incomes by over 600 percent. However, the task force found that the main challenges to increasing the use of small-scale water management methods are social and managerial, not technical, and places emphasis on the need to develop the social capital needed to ensure community action (United Nations 2004).

Agricultural water management has not received commensurate global attention.

Although international attention has been paid to both water resources management and to the hunger and food agenda, there has been little focus on the key questions that lie in the overlap between the two great global challenges of managing water resources sustainably and feeding the world. Producing ever more and better food for a world population growing at 80 million people a year takes water, for which there is increasing competition by domestic and industrial users and by the environment. More food

---

Box 3.1. The Challenge Program on Water for Food

The Challenge Program on Water For Food is an international, multi-institutional research program coordinated by IWMI, with a strong emphasis on north-south and south-south partnerships. The program supports institutions, research scientists, and development specialists in solving specific problems that occur when issues of water, food, environment, and poverty alleviation all meet. The program supports five interrelated thematic areas of research: (a) crop water productivity improvement; (b) water and people in catchments; (c) aquatic ecosystems and fisheries; (d) integrated water management systems; and (e) global and national food and water systems. The program takes a basin approach, on the view that the river basin is where the water problems and issues converge, especially in the developing world. Benchmark basins are those of the Yellow River, Mekong, Indus-Gangetic, Limpopo, Volta, Nile, Karkheh, Sao Francisco, and Andean basins.

Program activities seek answers to the question how to produce more food and sustain rural livelihoods with less water in a manner that is socially acceptable and environmentally sustainable. Answers are sought from two quarters: The first explores the food-related part of the challenge, examining issues of agricultural production, biology, physical science, and policy. The second focus is on resource management research at local, community, system, subbasin, basin, regional, and global levels.

Source: Personal communication, Pamela George, IWMI Program Manager, June 2005.

can certainly be produced for less water, but productivity improvements by their nature have impacts on the environment, on which society must agree. Cheap food is an almost universal policy goal, yet low food prices give little incentive to invest in efficient water management, and keep many food producers themselves in poverty. These issues have not yet been addressed by the global community in a coherent, integrated dialogue on AWM, food, livelihoods, and the environment.

Initial debate has sparked as much controversy as it has allayed. At the Second World Water Forum in The Hague in 2000, the possible trade-offs between water for food production and water for nature became one of the most contentious issues. Participants in the Water for Food theme stressed the need for continued—albeit slow—growth in water consumption in agriculture, while adherents to the Water for Nature theme called for significant reallocation of water from agriculture to environmental uses. The Global Water Partnership's (GUP) Framework for Action discusses the trade-off between the need to divert water from irrigated food production to other users and to protect the resource and the ecosystem as a source of potential conflict under conditions of growing water stress (Rosegrant, Cai, and Cline 2002b). Pioneering work by IFPRI, IWMI, FAO, and other agencies is now starting to bring the issues to the fore. Recent publications (Rosegrant, Cai, and Cline 2002a, 2002b; FAO 2003d; ADB/IWMI 2004) explore the water-for-food challenge, and are beginning to highlight some of the inherent dilemmas. In the future, AWM and its interface with food security, incomes and poverty reduction, and environmental sustainability have to become central topics for analysis and debate by the global community. A start was made by the formation of the Dialogue on Water, Food and Environment where some of the international actors[4] initiated a dialogue intended to develop a "science-based consensus" among all stakeholders—including governments, NGOs, research specialists, and farmers' organizations—on food security and poverty eradication in developing countries through the sustainable use of water resources. This program has, however, come to an end.

International research is more focused now than in the past on agricultural water management.

Internationally funded agricultural research is critically important in developing global public goods to reduce poverty and hunger (World Bank 2005a). This role is particularly important for AWM in the "more for less" agenda—more crops, jobs, and income per drop—crucial to improving irrigation productivity. International funding is also vital to compiling the list of incremental improvements that can increase soil moisture availability and return more per drop in low-yielding rainfed agriculture. The

Table 3.1. Research Programs and Projects of CGIAR Institutes Relevant to Agricultural Water Management

Research center	AWM programs and projects
ICRISAT	Under its Agroecosystems Global Theme, ICRISAT is researching low-cost AWM improvements that are risk-reducing and income-generating, together with improved policies for efficient water use and management, and community-participatory approaches to AWM.
ICARDA	ICARDA specializes in the issues of AWM in dry areas, including both rainfed farming and mitigation of drought. Under its Management of Scarce Water Resources and Mitigation of Drought in Dry Areas program, ICARDA is researching options for improving the productivity of water and for mitigating drought, including water resources management, drought tolerant and water-use efficient germplasm, and agronomic management of cropping systems. ICARDA is also researching the policy and institutional environments needed to support water-efficient technologies and drought-mitigating practices.
IRRI	IRRI specializes in productivity of rice cultivation. Its research focuses on natural resource management and water productivity under intensive rice systems. A parallel research program on productivity in fragile environments is examining AWM for rainfed rice ecosystems.
WARDA	WARDA's research program is focused on enhancing the performance of irrigated rice-based systems in Africa, including improved resource-use efficiency, options to mitigate environmental degradation, and improved lines and varieties for African irrigated rice-based systems.

Source: ICRISAT, ICARDA, IRRI, and WARDA Web sites (2005).

Consultative Group on Agricultural Research (CGIAR) has long identified the centrality of the AWM challenge to its work.

Among the 15 CGIAR research centers, the IWMI deals explicitly with AWM. IWMI research areas cover integrated water resource management for agriculture; sustainable smallholder land and water management systems; sustainable groundwater management; water resources institutions and policies; and water, health, and the environment (see box 3.1).

A number of the CGIAR crop research institutes include water management in their research agendas, and some also have a specific focus on the socioeconomic, policy, and institutional aspects of AWM. Four have a particular focus on AWM in their respective agro-climatic regions: the International Crops Research Institute for the Semi-Arid Tropics (ICRISAT), the International Center for Agricultural Research in the Dry Areas (ICARDA), the International Rice Research Institute (IRRI), and the Africa Rice Center (WARDA) (see table 3.1). Their research focuses on three key areas:

- Managing water in an integrated manner together with other natural resources (soil and so forth) and agricultural inputs (nutrients and pesticides) to achieve maximum productivity, risk mitigation, and water conservation and environmental protection
- Developing new crop varieties that are less susceptible to drought, floods, and salts; more productive per unit of water; less vulnerable to pests and disease; and less demanding of water-polluting fertilizers and pesticides
- Developing cropping systems and on-farm agronomic practices that are pro-poor, water saving, and socioeconomically viable, given markets and institutions

Another CGIAR institution, the International Food Policy Research Institute (IFPRI), focuses on institutional and economic research on policy solutions that cut hunger and malnutrition. IFPRI created and maintains a global modeling framework, IMPACT-WATER, which combines a model for policy analysis of agricultural commodities and trade with a basin-scale water simulation model. Directions for change for the global research agenda are discussed in chapter 5.

Agricultural trade agreements are affecting the incentive structure for irrigated agriculture.

Agriculture has a larger tradable component than many sectors and therefore is profoundly affected by the trade environment and trade policy (World Bank 2005a). This is especially true for some major irrigated commodities such rice, sugar, cotton, wheat, and the like, the returns to which depend on market-derived incentives. The global trade environment is thus of critical importance for irrigated agriculture in developing countries. The border prices of many important irrigated commodities from developing countries are depressed by developed countries' domestic subsidies, export subsidies, and especially import tariffs and restrictions. Organisation for Economic Co-operation and Development (OECD) subsidies (including the effects of trade measures) averaging US$238 billion annually (2001–3) have kept down the prices of sugar, cotton, and cereals to the point that several developing-country producers find exports uncompetitive and often protect their own production against imports from low-priced, developed-country production. Producer subsidy equivalents in developed countries are significant for commodities that are typically irrigated in many developing countries, such as wheat (40 percent), sugar (50 percent), and rice (as high as 80 percent).

In addition to facing depressed market prices, developing countries face restricted market access, with the persistence of bans, quotas, and tariffs on trade into many developed-country markets. Yet, where there are more favorable market access conditions, the dynamic impact of market-driven growth on irrigation development and productivity has been great. For example, horticultural products, the fastest-developing irrigated crops in developing countries (see chapter 1), enjoy an especially advantageous position, because market demand is rising and diversifying quickly and many developing countries enjoy a comparative advantage. These products are subject to relatively low tariffs in OECD countries, particularly for off-season production. Not surprisingly, competition is intense and quotas for horticultural products have become an important element in trade negotiations. Future development of the trade agenda in horticulture will be a significant driver of intensified irrigated agriculture in developing countries. The specific challenges of the trade and market development agenda for irrigated agriculture are discussed in detail in chapter 5.

3.2 CHANGING WATER RESOURCES MANAGEMENT PRIORITIES ARE AFFECTING AWM POLICIES.

The nature of the water resource constraint on AWM was described in chapter 2. With growing water scarcity, increased competition among sectors, and growing environmental concerns, decision makers in most countries face pressures to balance several not always easily compatible policy goals:

- Allocating water on a priority basis to domestic needs
- Developing equitable mechanisms for transferring water out of agriculture where needed
- Increasing the amount of water allocated to environmental uses
- Meeting quickly rising demand for agricultural products
- Raising rural incomes and reducing poverty

The weight assigned to each of these policy goals is different in each country and basin. However, in general, irrigated agriculture and AWM are facing two mounting challenges: to increase productivity in an environmentally friendly way so that food output and rural incomes will grow, but at the same time to surrender water—or at least forgo extra—in favor of domestic, industrial, and environmental needs. Resulting changes in AWM discussed below include the introduction of basin management approaches; a new emphasis on watershed management; the use of demand management instruments; new approaches on the supply side, including reuse of secondhand water; and management of climate change risk.

Agricultural water in the context of river basin management. Emerging best practice reviews agricultural water investments together with other uses within the river basin according to the contribution of each to overall basin water-use efficiency and water quality. Typically, basin management approaches are consultative and participatory, seeking to balance the views and goals of stakeholders and to coordinate among the many institutions involved in water management and water-related activities within a basin. They take into account not only technical feasibility and economic returns, but also environmental and social impacts and sustainability.[5] Basin management will consider, for example, how dams may reduce the downstream flow of water, how deep irrigation wells dry up traditional springs and shallow dug wells, or how agricultural drainage pollutes downstream water sources. The move toward these approaches is neither easy nor uniform. The Operations Evaluation Department (OED) of the World Bank (World Bank 2002a) was unable to find any World Bank–financed projects that systematically reviewed pre-project water uses to determine the effects of the project on water access and use by different socioeconomic groups. However, OED found that an increasing number of more recent Bank-financed projects do take these aspects into consideration.

Watershed and drainage management. The river basin approach reveals the need for greater upstream (drainage) investment and management in relation to AWM interventions. Where AWM interventions are viewed within the whole basin context, upstream issues of watershed management and dams, and downstream issues of drainage and environmental impacts become clear—leading to the recent emphasis on watershed management, in which the World Bank has invested almost US$1 billion since 1993. One notable success, China's Loess Plateau Watershed Rehabilitation Project, has substantially reduced local soil erosion by 20–30 million tons annually (and the total sediment load of the Yellow River by about 1 percent). It has also brought substantial socioeconomic benefits to marginal farmers on the plateau (World Bank 2002b, 2005b).

AWM demand management approaches. Water scarcity has led to more emphasis on demand management solutions as a means of ensuring that water is allocated to its highest-value use and is used efficiently in pursuit of a nation's social and economic goals. Typically, regulation and rationing have been the first demand management approaches to be adopted in large-scale irrigation schemes where water is scarce. However, volumetric measurement is really required for effective rationing, but in most irrigation schemes such measurement is difficult and costly to install. More recently, governments have seen the potential of participatory approaches to improve efficiency and contribute to water conservation, through their capacity to create ownership and responsibility, including group responsibility (see below). Education and information are also increasingly seen as important elements in any water conservation strategy. More contentious and politically difficult has been the use of the financial incentive structure: prices that reflect water scarcity can improve water allocation and encourage conservation, but raising prices has proved difficult for governments almost everywhere in the developing world. Other demand management instruments include the development of water rights and water markets. All these instruments are discussed in detail in chapter 5. In addition, the role of promotion and adoption of water conservation technology in managing demand is discussed in chapter 6.

Reuse of drainage and wastewater. Wastewater reuse in water-constrained countries is likely to become a major new water source. Reuse of water is already part of integrated water resources management policy in several countries, including Tunisia, Israel, Jordan, and the West Bank and Gaza, where water is scarce and the cost of developing new freshwater sources is high. However, countries trying these solutions are finding that they are far from risk free. Problems encountered include risks from the use of untreated wastewater to human health and to the environment, and problems of contaminants in drainage water, such as salt, metals, and pollutants. See chapter 6 for a full discussion of the issues.

Adapting to climate change. Climate is changing worldwide and there is some evidence of a growing trend of extreme climatic events. The impact of these changes is likely to be particularly injurious to developing countries, because they are mostly located in more-at-risk areas, are more dependent on vulnerable economic sectors such as agriculture, and have less capacity to adapt due to lack of resources. Effects of climate change on irrigated agriculture will be mostly driven by changes in water availability and quality on the one hand, and by changes in average and maximum temperatures on the other hand. In water-scarce regions, climate change is expected to further reduce both water availability—due to increased frequency of droughts, increased evaporation, and changes in patterns of rainfall and runoff—and water quality, through saltwater intrusion and sea surges. The effect of temperatures on crop productivity will be particularly felt in the tropics, where yields may decrease with even minimal changes in temperature. Extreme weather events will also affect crop yields

Table 3.2 summarizes possible climatic changes in the 21st century and their likely impacts on water resources and agriculture. Overall, climate change is expected to increase the existing vulnerability of farmers.

Table 3.2. Possible Climatic Changes in the 21st Century and Their Likely Impacts on Water Resources and Agriculture

Projected changes during the 21st century in extreme climate phenomena and their likelihood[a]	Projected impacts on water resources and agriculture
Higher maximum temperatures; more hot days and heat waves over nearly all land areas: *very likely*	Increased risk of damage to a number of crops
Higher (increasing) minimum temperatures; fewer cold days, frost days, and cold waves over nearly all land areas: *very likely*	Decreased risk of damage to a number of crops and increased risk to others
Increased summer drying over most mid-latitude continental interiors and associated risk of drought: *Likely*	Decreased crop yields Decreased water resource quantity and quality
More intense precipitation events: *very likely over many areas*	Increased flood, landslide, avalanche, and mudslide damage Increased soil erosion Increased flood runoff could increase recharge of some floodplain aquifers

Source: Adapted from IPCC 2001.

a. "Likelihood" refers to judgmental estimates of confidence: very likely (90–99 percent chance); likely (66–90 percent chance).

Climate change introduces a risk factor into the hydrological assessment. The effects of climate change on irrigation demand are expected to vary widely in different geographical areas. Hydrogeological models developed in the last few years give different results. As an example, Döll and Siebert (2000, 2002) predict that net irrigation requirements would decrease across much of the Middle East and North Africa as a result of increased precipitation, whereas most irrigated areas in India would require more water. The extra irrigation requirements per unit area in most parts of China would be small, while there would be a greater increase in northern China. Other climate models would give different indications of regional changes in irrigation requirements. On the global scale, different climate models are more consistent with each other, and predict that global net irrigation requirements would increase, relative to the situation without climate change, by 3.5–5.0 percent by 2025 and 6–8 percent by 2075 (IPCC 2001).

Some countries, mainly in Sub-Saharan Africa, are likely to become more vulnerable to food insecurity, and climate change could increase the dependence of some countries on food imports. Overall, climate change will lower

Table 3.3. Regional Impacts of Climate Change

Region	Expected impacts in water, agriculture, and food security
Africa	Increase in droughts, floods, and other extreme events would add to stress on water resources and food security, constraining development. Changes in rainfall and intensified land use would exacerbate the desertification process (particularly in the Western Sahel and Northern and Southern Africa). Sea level rise would affect flooding, as well as increase the risk of saline water intrusion into aquifers, especially along the eastern coast of southern Africa. Major rivers, highly sensitive to climate variations, may experience decreases in runoff and water availability, affecting agriculture and hydropower systems, which may increase cross-boundary tensions.
Asia	Extreme events have increased in temperate Asia, including floods, droughts, forest fires, and tropical cyclones. Thermal and water stress, flood, drought, and tropical cyclones would diminish food security in countries of arid, tropical, and temperate Asia. Agriculture would expand and productivity would increase in northern areas. Reduced soil moisture in the summer may increase land degradation and desertification.
Latin America	Loss and retreat of glaciers would adversely impact runoff and water supply in areas where snowmelt is an important water resource. Floods and droughts would increase in frequency, and lead to poorer water quality in some areas. Increases in the intensity of tropical cyclones would increase the risk of damage to crops from heavy rain, flooding, storm surges, and wind damage.

Source: Adapted from Asian Development Bank et al 2004.

the incomes of vulnerable populations and increase the absolute number of people at risk of hunger. By contrast, production could be boosted in developed countries, so the location of world food production may shift, increasing the gap in trade and incomes between the developed world and the poorest countries, particularly those in Sub-Saharan Africa (IPCC 2001). See table 3.3.

Solutions, which countries are already starting to implement (see World Bank 2005b), include policies and plans for preparedness, the adoption of drought-tolerant crops, raising awareness at the farm level, early warning, and so forth. Policy and investment options for tackling the threat of climate change are discussed in chapter 5.

3.3 DEVELOPMENT APPROACHES IN AGRICULTURAL WATER MANAGEMENT ARE EVOLVING.

This section reviews how approaches to agricultural water management are changing globally, with aspects such as the environment becoming increasingly important.

International agencies, civil society, and governments are changing their approaches in critical areas of agricultural water management.

Attitudes and approaches among the global community to AWM issues are starting to evolve. At the World Bank, the three definitive corporate strategies on Water, Rural Development, and Environment, which were discussed in the Executive Summary to this book, reflect new shifts in thinking about AWM toward an emphasis on water productivity increases; toward a focus on institutions, governance, and poverty reduction; and toward integrated approaches that factor in environmental and social considerations.

International investment in AWM has moved toward system upgrading and management improvement. The character of international investment in AWM in developing countries has been changing over time. Until the early 1970s, the emphasis was on developing new infrastructure. Subsequently, there was a progressive shift to rehabilitation associated with implementation of management, institutional, and policy reforms. About two-thirds of recent international financing for irrigation and drainage—and almost all World Bank lending in the sector in the last decade—has been for rehabilitation and upgrading of large-scale irrigation. Most recently, investment in irrigation and drainage has taken place in the context of broader integrated water resources management approaches.

Few projects have addressed the central objective of achieving a demand-responsive water delivery service. Much of the investment in the 1990s addressed the huge backlog in deferred maintenance and repairs, supported by related improvements in institutions and management. Few projects tackled the challenge of integrated system modernization, that is, to change the irrigation delivery system and institutional and incentive structures to provide a sustainable, efficient, and demand-responsive water delivery service. Only a few success stories addressed this fundamental issue of focusing on improved water service, including the Office du Niger reforms in Mali and the large-scale Tarim II project in China. Modernization of service through upgrading existing infrastructure is a challenging technical and institutional problem, as illustrated by the Irrigation Improvement

Project in the Arab Republic of Egypt. Many years after developing the project concept, implementation of continuous flow in branch canals to replace a rigid rotation system started only recently.

These changes can be understood within the analytical framework for water sector investment that the Water Resources Sector Strategy of the World Bank maps out. This framework traces three stages in a common progression of water resources development in relation to investment. In a first stage, there are abundant water resources, and high returns to infrastructure, which therefore is the predominant area for investment. In the second stage, there are some still unharnessed water resources, but the country is experiencing water shortages. More investment is made in infrastructure rehabilitation and improvement and demand management instruments are introduced. In the final stage, the priority is management of scarce water resources and of existing infrastructure, and greater attention is needed to integrated basin management and watershed management, and to pollution control activities. Demand management becomes more important than supply management[6] (see figure 3.1).

The shift in the approach of international agencies reflects the fact that in many countries the irrigation sector is in the second or third stage of the investment progression, with most or all resources harnessed, and investment in irrigated farming is increasingly in intensification rather than in area expansion. This trend is likely to continue, with more than three-quarters of the increase in production in the next 25 years projected to come from modernization and intensification (FAO 2003d). International agencies have also increasingly financed investments in small-scale irrigation where poverty-reduction impacts may be greater. Examples include recent World Bank–financed projects in small-scale irrigation in Morocco, and in smallholder private irrigation in Niger, Burkina Faso, and Mali. At the same

Figure 3.1. Rates of Return on Investment by Stage of Development of Water Infrastructure

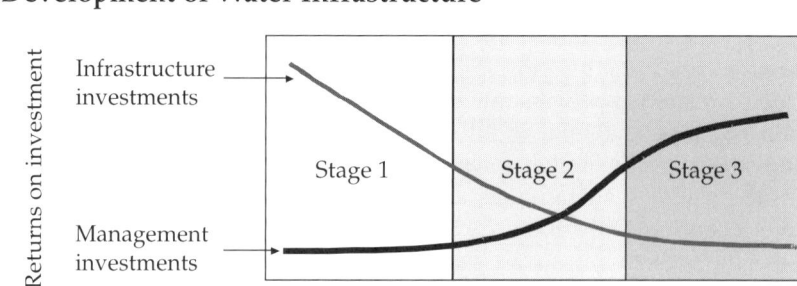

Source: World Bank 2002b.

time, there is awareness that some countries and river basins still have a significant potential for first-stage investments in new diversion and irrigation infrastructure. Ethiopia, for example, has developed only about 170,000 ha of its estimated irrigation potential of 2–3 million ha. As an extremely poor and populous country, but with abundant water resources (85 percent of Nile Basin resources), Ethiopia now sees new irrigation infrastructure development as a vital component of its economic development and poverty-reduction strategy (World Bank forthcoming). The Water Resources Sector Strategy reflects this awareness with its call to reengage with investment in infrastructure.

An increasing body of research shows that AWM can contribute to poverty alleviation in many developing countries. In allocating resources among countries and sectors, international agencies have increasingly emphasized poverty-reduction objectives. Within the irrigation and drainage portfolio, OED (World Bank 2002a) records that the poverty focus of World Bank investments increased by 23 percent after the Bank Water Resources Policy of 1993. OED's ongoing review of the World Bank irrigation and drainage portfolio is focusing particularly on poverty-related aspects. When the report is available, it will give a more precise evaluation of pro-poor impacts, which will be key in repositioning irrigation within poverty-reduction strategies and related national and donor-financed investment programs.

Consideration of the environment is also becoming an important factor in agricultural water management.

Globally, there have been changes in knowledge, attitudes, and politics regarding AWM and the treatment of the environment. Ecosystems are more highly valued. Understanding has grown steadily that the environment is a water-using sector, but that its uses are different and its constituency scattered. It has been called "voiceless," although its voice has been found in recent years, largely through NGOs and other representatives of stakeholders and civil society. There is broader understanding of who the stakeholders are in the environment; and there is generally more commitment to considering their needs, joined to growing social pressure for agriculture that is less harmful to the environment (FAO 2003e). Debate on the environmental impacts of irrigation has made it clear that irrigation cannot be neutral to the environment. There is a broader understanding of the multifunctionality of water and of human and ecosystem interactions. Some specific areas of interaction, such as environmental flows and non-point-source pollution, are much better understood (World Bank 2003c). The world, thus informed, is better armed to assess the inevitable trade-offs between the environment and increasing irrigated agricultural production. Some development initiatives have sprung from this awareness. Analytic and policy tools and technologies have been developed over the last 20 years that at least allow trade-offs to be faced in the light of knowledge of the risks and costs and of the mitigation measures to be developed.

One facet is the need to maintain or increase environmental flows. Environmental flows—the waters left in a river ecosystem, or released into it—are critical for maintaining ecosystems. The recognition that modifications to river flows are an important source of riverine, floodplain, and in some cases estuarine degradation is relatively recent. However, over the last 20 years methodologies have been developed, and the results are now starting to be integrated into overall basin planning frameworks (World Bank 2003c). (See also chapter 5.)

These environmental concerns are increasingly mainstreamed into international development business in AWM. Within the World Bank, safeguard policies (see below and chapter 6) have been developed to evaluate likely impacts, identify alternatives, and provide plans that minimize or mitigate harmful effects (World Bank 2002a). The Bank's environmental assessment safeguard policy (Operational Policy 4.01) is now triggered if modifications to river flows lead to adverse environmental risks and impacts. The Bank has also addressed environmental concerns regarding water directly through investments. Environmental projects are now the third largest category of water-related institutional support and investment in the Bank's portfolio. By 2002, there were 48 water-related environmental projects in the portfolio, with US$3.3 billion committed since 1993.

Environmental and social safeguards have been increasingly used to ensure that best practice is incorporated into irrigation investment projects. To ensure that environmental and social concerns are systematically reflected in the design of irrigation interventions, international agencies have increasingly codified best practice approaches into their way of working. Within the Bank, "safeguard policies" have been developed. The objective of these policies is to improve the quality of and the ability to implement projects, to minimize harm, and to ensure that, where there are trade-offs, they are dealt with in a fair and socially acceptable way. Seven of these safeguard policies apply particularly to irrigation and AWM (see box 3.2).

Experience of the application of safeguards has been, on the whole, positive (World Bank 2002a). The compliance at entry of water projects with safeguards is higher than that of other Bank projects. Since 1993, almost three-quarters of all water projects have been classified as environmentally sensitive, reflecting the identification of risks and the need to mitigate negative impacts. The safeguards procedures have routinely identified AWM projects involving involuntary resettlement, and the number of water projects that include resettlement has fallen, suggesting improved project selection and the identification of superior alternatives.

3.4 THE ROLES OF THE RESPECTIVE AWM STAKEHOLDERS ARE CHANGING.

This section looks at the institutional structure of AWM and the roles of different stakeholders, examining how those roles have changed over time, and future perspectives and trends.

The role of government in AWM is changing, but very slowly.

Governments have long been predominant in large-scale irrigation (which represents about half of all irrigated areas in developing countries). Chapter 2 discussed the outcomes of many years of virtual state monopoly of investment and management in the large-scale irrigation sector: high fiscal cost, suboptimal system efficiency, and production and income shortfalls. By the 1990s, most development agencies were actively advocating reforms in the irrigation sector, emphasizing a reduced role for the government and a larger one for the users, financial autonomy for irrigation agencies, and devolution of management responsibilities to water users' associations, at least at the lower levels of schemes.

This movement was supported by broader new thinking in the OECD countries about the role of government. New approaches to public management that emerged in the 1980s began to define a minimum essential role for the state and increased roles for the private sector and civil society. These approaches perceive government not as producer and manager, but as regulator and promoter of the entrepreneurial energies of the people,

---

Box 3.2. World Bank Safeguard Policies

Seven safeguard policies apply particularly to AWM operations:
Environment
- Environmental assessment (OP 4.01)
- Natural habitats (OP 4.04)
- Safety of dams (OP 4.37)

Social development
- Cultural property (OPN 11.03)
- Involuntary resettlement (OP 4.30)
- Indigenous peoples (OP 4.20)

International law
- Projects on international waterways (OP 7.50)

Source: Adapted from World Bank 2002a.

with the state entering into a direct managerial or subsidy role only when social justice demands. These models were implemented widely in OECD countries during the 1980s and 1990s.

However, in practice, only a few developing countries have made progress in implementing these changes. Reform in the irrigation and drainage sector has been slow, in part because many of the models required untested institutional reorganizations, which represented radical departures from the historic bureaucratic, top-down approaches to large-scale irrigation. Political economy considerations have been an important brake on these changes, which carry high political transactions costs (see chapter 5 for a discussion of the political economy of reform). In most developing countries, particularly in Asia, which accounts for two-thirds of the developing world's irrigated area, centralized planning approaches are still dominant in irrigation and drainage development and the dominant model for large-scale irrigation remains a government run and largely subsidized scheme. Although some movement toward scheme financial autonomy and full operations and maintenance cost recovery has been initiated, only a few large-scale public irrigation schemes (in China and Tunisia, for example) have become financially self-sustaining to cover operations and maintenance expenditures. Cost recovery generally remains low. In some cases, institutional reforms have actually been unsuccessful. In the late 1980s in Madagascar, for example, regional irrigation public sector agencies were dismantled, publicly managed schemes were hastily transferred to user responsibility, and government budget support was almost totally suppressed. Without proper preparation and continuing support, public schemes covering almost 100,000 ha virtually collapsed.

The most significant change in institutional arrangements has been the participatory irrigation management (PIM) movement.

The division between public and private is being rewritten, largely through decentralization and user participation processes. Decentralization takes several forms in AWM: delegation of service provision functions to locally autonomous public bodies or to stakeholder organizations; involvement of users in planning and managing water projects; or handover of schemes to user organizations or a management company. The movement toward decentralization is reflected in government investment patterns. In recent years, about 70 percent of World-Bank-financed water projects addressed decentralization of water resource management. OED (World Bank 2002a) found that approaches in Bank-financed irrigation and drainage projects in the last decade were generally participatory, incorporating the views of stakeholders in project design and establishing social impact and poverty monitoring. The role of women in water has been increasingly considered—OED (World Bank 2002a) found that 54 percent of World Bank water projects after 1993 addressed gender, compared to 30 percent previously.[7]

The most significant change in institutional arrangements in recent years has been the participatory irrigation management (PIM) movement and the development of water user associations (WUAs). As a counterpart to the redefinition of the role of public institutions, WUAs have developed over the last decade as a way to decentralize management and involve stakeholders responsibly. The underlying rationale for participation in irrigation is that users have a direct interest in the efficiency and flexibility of water delivery because of its influence on profitability. Users are more willing to pay for costs if they have an influence over operations. By the mid-1980s, several countries were testing participation in operation and maintenance activities. In the Philippines, Indonesia, and Pakistan, these programs consisted of a large-scale transfer of the lower levels of canal irrigation systems to user groups, each one covering a few hundred ha (50–200 ha). In the early 1990s, Mexico undertook a more ambitious, two-phase transfer program. In the first phase, now completed, user associations took over the financial and managerial responsibilities for operating systems below the main canals, areas ranging from 5,000 to 25,000 hectares. In the second phase, responsibility for managing the main canals is being handed over to limited liability companies. The success of the transfer program in Mexico has encouraged other countries, such as Turkey, to adopt the same approach, with similar success.

WUAs operate now in over 50 countries, involved in operation and maintenance, setting and collecting fees, hiring professional staff, managing water rights, and so forth. They have proven, in the best cases, to be efficient, accountable, and responsive—but not in all cases. Associations have been much more successful than government agencies in recovering costs through higher charges and higher collection rates. Maintenance activities by the associations have helped stop the deterioration of infrastructure, but the impact of WUAs on efficiency and productivity is mixed. Overall, experience shows that participatory approaches, properly undertaken, can reduce costs to government and improve scheme management. Essentially, these changes work best when physical and institutional improvements are implemented in a coordinated manner (Vermillion 2004).[8]

Scope for farmer and other private sector investment is increasing.

Privately managed schemes cover the larger part of the total irrigated area—private groundwater irrigation alone accounts for over half of irrigated area worldwide, and private investment for half of total investment (see

---

Box 3.3. Private Investment in Irrigation in Latin America

In Latin America, private sector investment in surface irrigation has historically been important. In Mexico, around 40 percent of the irrigated area was privately owned, even before reforms to publicly funded irrigation districts started to shift control to water user associations in the early 1990s. Following the reforms, increases in private sector investment in irrigation infrastructure in Mexico have been dramatic. Government has reduced public investment in irrigation substantially (by more than 40 percent between 1991 and 1995).

In Chile, with one of the most privatized irrigation sectors in Latin America, farmers must, by law, contribute as much as 75 percent to new pumping and channel irrigation projects, with the result that only the most profitable schemes are built. Private sector involvement in the approval, funding, operation, management, and maintenance of irrigation projects has increased water efficiency and contributed to the boom in agricultural exports.

Source: Lipton and and others 2005.

---

chapter 6). Historically, private investment has been the rule in small-scale irrigation, and worldwide farmers have invested in groundwater extraction, which has been by far the fastest growing AWM activity in recent years. In India and Mexico, two-thirds of groundwater development has been financed entirely by the private sector. In some countries, particularly in Latin America, private sector investment in irrigation is dominant and has improved efficiency (see box 3.3). On some large-scale surface schemes, private contractors provide services for a fee. For example, in Shaanxi Province in China, contractors operate a local irrigation system based on a multiyear contract signed between the contractors and users.

Over the last decades, governments have begun to coinvest with farmers in small-scale irrigation, in watershed management, and in supplementary irrigation. These projects have often matched public investment criteria better than large-scale irrigation, because they are not only higher return (to capital and to water) but also more pro-poor. More recently, there have been promising pilot projects of public-private partnership in irrigation investment and management, the most recent case being the Morocco Guerdane private sector build-operate-transfer contract, which was awarded in July 2004 (see chapter 6).

4

The Future Contributions of Agricultural Water Management and Potential Risks

Chapter 1 reviewed the way in which the production of food and other commodities has been able to meet quickly rising demand, largely through improved productivity, which accounted for over three-quarters of the increase in agricultural production in developing countries over the last 30 years. Irrigation and improved agricultural water management (AWM) played the key role in this extraordinary achievement, with water productivity doubling globally over the last four decades. For many crops, the spectacular yield increases over this period, were, in fact, achieved without any increase in water consumption.

In light of the vectors of change discussed in chapter 3, the present chapter examines the scope and nature of the likely increase in demand for agricultural products in the coming years and the technical capability of irrigation and AWM to respond. The supply side discussion covers the scope for further increases in water productivity and cropping intensities, and whether new water withdrawals for irrigated agriculture and extension of the irrigated area are feasible. The chapter also examines the considerable risks involved in meeting the challenge of rising demand, and outlines what changes may be required to manage the needed growth in a socially and environmentally acceptable way.

4.1 MATCHING FUTURE SUPPLY AND DEMAND FOR AGRICULTURAL PRODUCTS WILL CONTINUE TO BE A CHALLENGE FOR AGRICULTURAL WATER MANAGEMENT.

Strong population growth, higher GDP growth rates, and increased incomes are likely to increase both demand and access to food.[9] Globally, the world food challenge will remain enormous. World population is expected to increase by 40 percent between 1999 and 2030, with an average of 80 million new mouths to feed each year (FAO 2003a). Developing-country populations are projected to increase by half over the period. The GDP of developing countries as a whole is expected to grow by an annual average of 4 percent over the next three decades. Expected high per capita income growth rates of over 4 percent for South and East Asia mean that for these regions as a whole, the specter of household food insecurity should dwindle rapidly. Low anticipated per capita income growth rates of only 2 percent a year in Sub-Saharan Africa reflect the likelihood of both slower GDP growth and higher population growth rates in that region. With potentially low levels of household income, the population in Sub-Saharan Africa will remain vulnerable to food insecurity.

Food self-sufficiency rates will decline, with varying impacts on country AWM strategies. High economic growth rates for regional economies will allow food deficit countries in South Asia, East Asia, and Latin America to import an increasing share of their basic food needs. This will stimulate investment in higher-value irrigated agriculture where markets exist. Where countries decide to import more, pressures to intensify cereals production further will ease and may allow the release of some agricultural water for domestic and industrial purposes, or for the restoration of environmental flows. By contrast, low levels of GDP in Sub-Saharan Africa and low projected growth rates will make it hard to import more food. If nothing changes, per capita food consumption will remain the lowest in the world, with consequent likely risks to nutrition and hunger. Agricultural development in Sub-Saharan Africa is likely to focus on strategies to improve local food crop production in currently subsistence environments and to develop irrigated agriculture where investment costs are not too high. AWM will be an essential element in both strategies.

Meeting food demand will place great strain on irrigated production systems and the resource base.

Recent projections highlight the magnitude of the challenge for AWM. Two recent exercises estimate the supply changes needed to meet the expected rise in demand (FAO 2003d; IFPRI/IWMI in Rosegrant, Cai, and Cline 2002b).[10] Both analyses conclude that to meet the increased demand will require continued increases in water productivity, cropped area, and water withdrawals, although at a much slower rate of increase than in the last 30 years. The main variables used by the two analyses are broadly similar, and are summarized in table 4.1. (All variables relate only to developing countries.)

To meet demand, FAO estimated that crop production in the developing world would need to increase at about 1.6 percent per year over the next three decades (see table 4.2). The highest expectations are in Sub-Saharan Africa, which alone of all regions would have to register higher rates of annual increase in crop production than in recent years. Again, this focuses attention onto the need to find ways to accelerate agricultural growth in that region. For developing countries as a whole, public policy—both international and national—will have to focus on the investment, technology, and incentive packages needed to prompt such a rapid rate of growth.

Table 4.1. Summary of Selected Variables in FAO and IFPRI/IWMI Supply and Demand Projections for Developing Country Irrigated and Rainfed Crop Production

Indicator	Base year value (1997–9 unless specified)	Future value (2030 unless specified)	Increase (percent)
Arable land in production			
total irrigated land	202 M ha	242 M ha	20
total arable land	956 M ha	1,076 M ha	13
Irrigation water withdrawals	2,128 Bcm	2,420 Bcm	14
Cropping intensity			
irrigated	127%	141%	11
rainfed	83%	87%	5
Irrigation efficiency	38%	42%	11
Yields			
irrigated cereals	3.25 t/ha (1995)	4.60 t/ha (2025)	42
rainfed cereals	1.50 t/ha (1995)	2.13 t/ha (2025)	42

Source: All figures are from FAO 2003d, except figures for 1995 and 2025, which are from IFPRI/IWMI in Rosegrant, Cai, and Cline 2002b.

Notes: M ha = millions of hectares; Bcm = billion cubic meters; t/ha = tons per hectare.

Table 4.2. Annual Percentage Rates of Increase in Crop Production Projected by Region of the Developing World, 1969–2030

Region	1969 to 1999	1997–9 to 2030
All developing countries	**3.1**	**1.6**
East Asia	3.6	1.2
Latin America	2.6	1.7
Near East and North Africa	2.9	1.6
South Asia	2.8	1.8
Sub-Saharan Africa	2.3	2.5

Source: FAO 2003d.

Irrigation and improved water management will be core features of those packages, because they are essential to high agricultural growth rates.

Irrigated areas are likely to provide more than half of the increased crop production.

FAO projects that output of cereals in developing countries would have to rise by 61 percent from 1997–9 to 2030, an extra 626 million tons. Output of other largely irrigated crops is projected to grow faster still—sugar cane

by 70 percent, cotton production by 90 percent. Production of maize is projected to double. Overall, irrigated agriculture would bring 57 percent of the
total increase in crop production.

Yield increases, largely stemming from irrigated agriculture, would have
to account for two-thirds of increased output—and for 80 percent in land-
constrained Asia (see figure 4.1).

Irrigated yields for all crops will need to increase by about half on average. For cereals, FAO projects that irrigated yields in developing countries
need to increase from the present 3.2–3.5 tons/ha to 4.5–5.0 tons/ha, with
rice yields going up by 25 percent, and wheat yields by 30 percent (bringing average wheat yields in 2030 to four times the level at the start of the
1960s) (figure 4.2). By 2030, average cereals yields in developing countries
will have to be higher than cereals yields in the developed world today, a
considerable challenge.

Food demand will not only increase, but also change in composition, with
consequences for water demand. It is likely that the current trends in developing countries toward a broader and more diversified diet will continue, and
that consumers will show increasing preference for safe, high-quality food.

The pattern of improving diet in recent years will continue (table 4.3)
and is expected to place increasing demands on agricultural water resources.

**Figure 4.1. Anticipated Sources of Growth in Crop Production,
1997–2030**

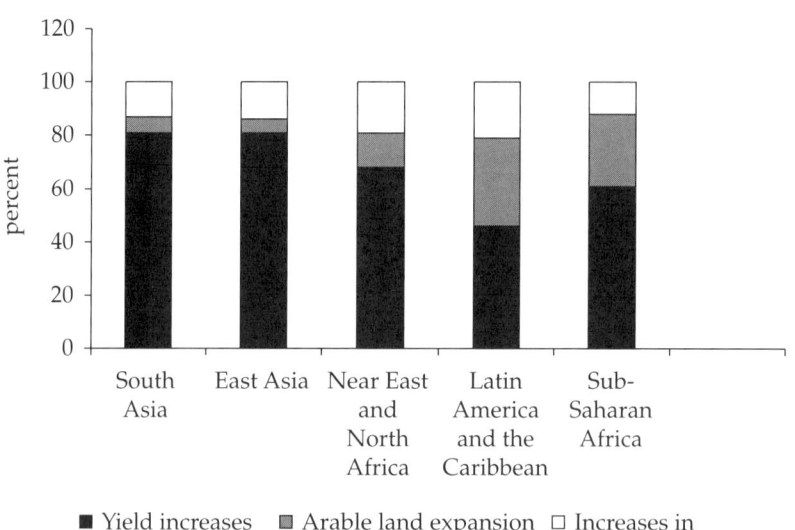

Source: FAO 2003d.

Direct consumption of cereals per person is not expected to change, but more cereals will be processed into animal feed and other uses. Increasing quantities of the most water-intensive agricultural product—meat—will be consumed. However, diets will likely continue to shift from beef (with its poor conversion rate of cereals to meat weight of between 5:1 and 7:1)

Figure 4.2. Projected Increases in Production and Yields for Predominantly Irrigated Crops in Developing Countries
(1961–3 = 100)

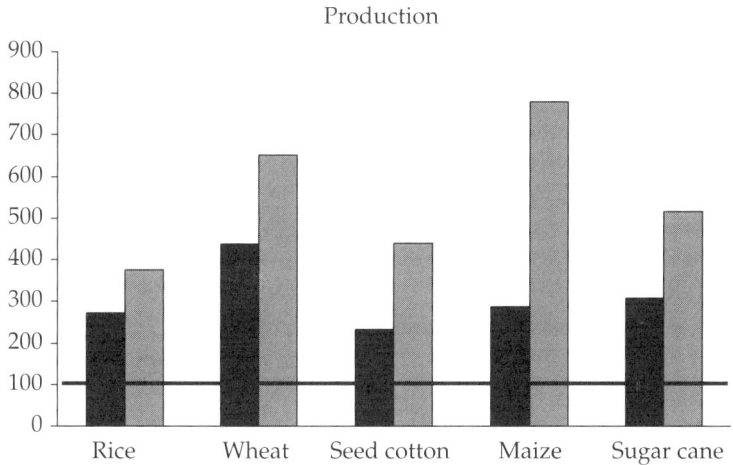

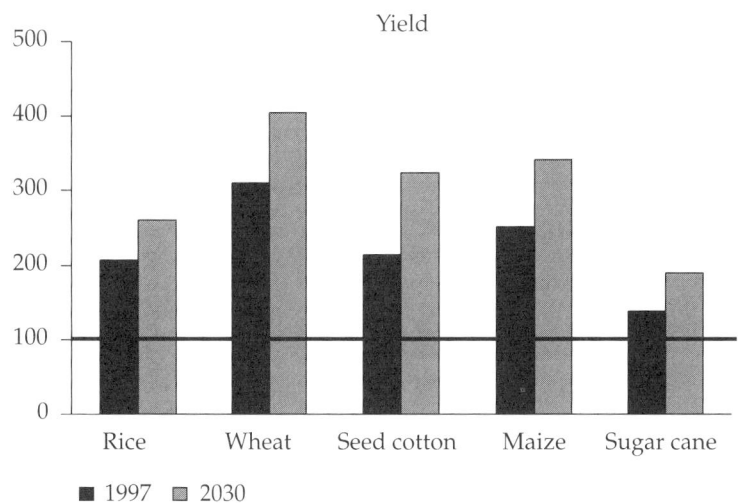

Source: FAO 2003d.

Table 4.3. Projected Changes in the Commodity Composition of Food Consumption for 93 Developing Countries
(kg per person per year)

Commodity	1997–9	2030
Cereals (food)	173	172
Cereals (feed and other uses)	74	107
Roots and tubers	67	75
Sugar	21	25
Pulses	7	7
Vegetable oils	10	15
Meat	26	37
Milk and dairy	45	66
Total calories per day	2,681	2,980

Source: FAO 2003d.

to poultry (a much more efficient and less water-intensive food, with a 2:1 conversion rate). Demand for fresh fruit and vegetables is likely to continue to be strong. Although these crops are more water-intensive than cereals, their high value per unit of water used makes them more "water productive" (FAO 2003d).

The challenge will be greatest in South Asia and Sub-Saharan Africa, the world's two poorest regions. In South Asia, relatively high rates of increase in production will be required, averaging 1.8 percent a year, yet water and land resources are already intensively exploited—only 6 percent of increased production in the region is expected to come from expansion of arable land area. In Sub-Saharan Africa, where the projected rate of increase in production is the highest of all regions at 2.5 percent, production systems are largely traditional, and the challenge of improving AWM is the greatest. Here, increases in the harvested area (through investment in irrigation infrastructure, land expansion, and improvements in cropping intensity) will be important factors, but improved AWM, inputs, and husbandry will be required to generate the more than 60 percent of increase projected to come from productivity improvements.

4.2 AS DEMAND FOR IRRIGATED CROPS GROWS AND WATER AND LAND RESOURCES ARE CONSTRAINED, WATER PRODUCTIVITY MUST INCREASE.

As demand grows and supply is constrained, improved water productivity will be essential. The water productivity achievement of recent years has been extraordinary: the water needed to feed a person has halved in the last 40 years (from 6 m^3 to 3 m^3 a day, see chapter 1). With irrigated production challenged to increase by two-thirds over the next 30 years while using little more water (14 percent more, according to FAO 2003d), there will be strong pressure to increase water productivity further. Emphasis will be on three components of water productivity: (a) increasing irrigation efficiency to convey water to the plant root more efficiently; (b) improving yields per cubic meter of water consumed; and (c) managing cropping patterns, input costs, and marketing to increase income and employment per cubic meter of water consumed. See box 4.1.

Box 4.1. Defining Irrigation Efficiency, Crop Water Productivity, and Evapotranspiration

Irrigation efficiency is the ratio between water withdrawn and water beneficially used by plant roots. It can be measured at the farm, scheme, or basin level. *Crop water productivity* is defined as the marketable crop output (in production or income) over consumptive water use. Consumptive water use refers not to water withdrawals from surface or groundwater but to the portion of water withdrawn that is *depleted,* that is, lost to the system and not returned or recycled to the groundwater or surface water system.

This depletion or *evapotranspiration* (ET) takes into account all of the evaporation and transpiration, both beneficial and nonbeneficial, in a given area. Thus, crop water productivity may be calculated either as unit of product per unit of ET (kg/m^3), or as net farmer income per unit of ET ($/m^3). In water-scarce areas or where there is overexploitation, real water savings are achieved through the reduction of ET. ET reduction can be achieved through an integrated set of engineering, agricultural, and management measures, and irrigation efficiency improvements should be analyzed in this light.

Crop water productivity shows large ranges even in comparable agro-climatic and production situations. Using the kg/m^3 ET measure from 82 literature sources, Zwart and Bastiaanssen (2005) found the range for wheat to be 0.5–1.9 kg/m^3; for rice 0.5–1.7 kg/m^3; for seed cotton 0.39–0.95 kg/m^3; and for maize 1.0–3.0 kg/m^3. Using the $/m^3 ET, the International Water Management Institute analyzed water productivity data for a total of 23 irrigation systems in 11 countries in Asia, Africa, and Latin America: values ranged from US$.03 per m^3 (for a system in India) to US$.91 per m^3 (for one in Burkina Faso), with an overall average of US$.25 per m^3.

Sources: Personal communication from Douglas Olson , World Bank, March 2004; Zwart and Bastiaanssen 2005.

Irrigation efficiency can be improved.

Irrigation efficiencies will have to rise, especially in water-scarce regions. FAO (2003d) projects further increases in irrigation efficiency: from 38 percent in 1997–9 up to 42 percent in 2030 (see figure 4.3).[11] Higher than average irrigation efficiencies are generally found in water-scarce regions such as the Middle East and North Africa (average over 40 percent) or the north of China. Low efficiencies generally tend to persist in situations without water stress. There is certainly potential in most regions for further increases in efficiency.

A wide range of known investment, institutional, and water management improvements can enhance irrigation efficiency, including physical improvements (irrigation system modernization, land leveling, selective canal lining, and pressurized irrigation, for example), institutional changes (improved management practices), and demand management incentives to reduce on-farm water wastage (pricing, deficit irrigation, and the like).

System modernization has been widely used to improve large-scale irrigation efficiency. Modernization programs include a broad range of hardware and software improvements (see chapter 6), all designed to improve the efficiency and timeliness of water service to the farmer: for example, adapting an open-channel irrigation system suitable for a field crops monoculture to pressurized pipes to allow farmers to diversify. Improved water service is especially important where farmers are producing higher-value

Figure 4.3. Irrigation Efficiencies, 1997–9 and 2030

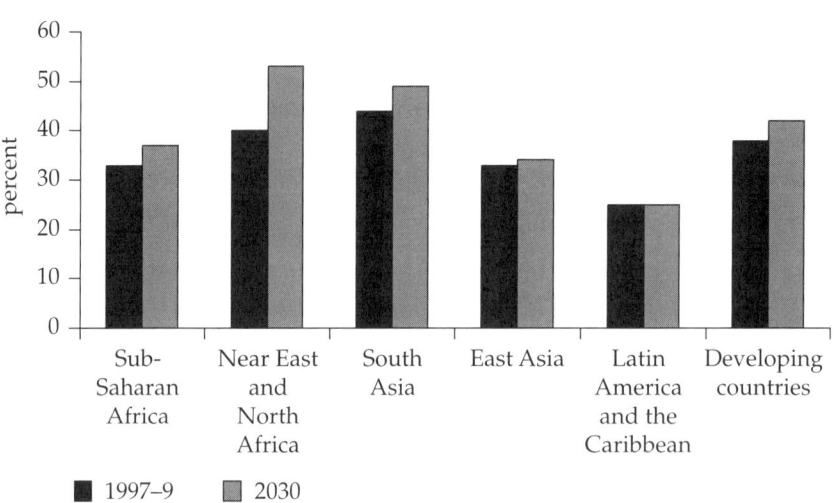

Source: FAO 2003d.

crops for market, because such crops need assured water delivery with controlled flow rates delivered directly to the plot, together with good drainage. The resulting irrigation efficiencies can be high indeed: efficiencies of 70 percent and higher are regularly attained on schemes in California. After poor experience in modernization programs that incorporate only "standard" solutions such as canal lining, recent techniques for planning modernization programs focus on finding the most cost-effective ways of ensuring timely water delivery to the plant (figure 4.4). As irrigation technology becomes more sophisticated and high-value crops are grown, a higher level of farmer management skill is required.

The use of groundwater and pressurized irrigation is more efficient than conventional surface irrigation. Groundwater is controlled by the farmer, who can irrigate virtually on demand, and it is continuously available, even during droughts. Groundwater is usually available close to the point of use, so conveyance costs and losses are reduced. Groundwater also offers natural storage capacity, reduced evaporation losses, and improved drainage. Where groundwater is conveyed through piped and drip systems, its efficiency is even greater (and water productivity is often two to three times higher than that of surface irrigation—see table 4.4). Scope for further improvements in irrigation efficiency with groundwater does exist, particularly in adoption of pressurized irrigation. (There is also scope for improving productivity through cropping patterns and husbandry practices—see below.)

Even the higher cost of groundwater brings advantages in incentives to efficient use and to adopting higher-value cropping patterns. The technology is appropriate on a wide scale. Less than 1 percent of the irrigated area of India, for example, is under pressurized irrigation, but up to half of the irrigated area is potentially suitable. However, as described in chapter 2,

Figure 4.4. Potential Efficiency of Alternative Irrigation Systems

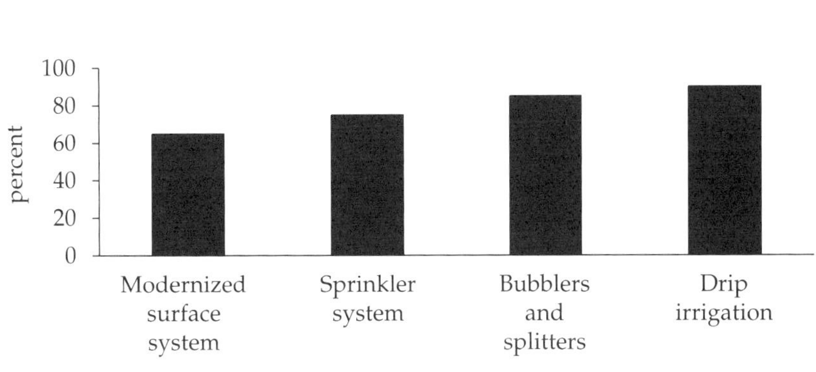

Source: Gleick 1993.

Table 4.4. More from Less: Water Productivity Gains from Shifting to Drip from Conventional Surface Irrigation in India

Crop	Change in yield/ha (percent)	Change in water use/ha (percent)	Change in water productivity (percent)
Bananas	+52	–45	+173
Cotton	+27	–53	+169
Grapes	+23	–48	+134
Sweet potatoes	+39	–60	+243
Tomatoes	+50	–39	+145

Source: Postel 1999.

the resource constraint for groundwater is particularly acute, with the likelihood that in many basins, groundwater use would have to decline to achieve sustainability. Efficiency can be further enhanced when groundwater is used in association with surface water (conjunctive use) or as a supplement to rainfall.

In all "efficiency" improvements, care has to be taken, when the resource is scarce and opportunity costs are high, that there are real water savings. Of the total amount of water withdrawn, a portion is depleted and a portion is returned or recycled to the groundwater or surface water system. Improving irrigation efficiency often results in an increase in the depletion fraction and a decrease in the return flows, and hence in increases in water consumption. Where water is abundant this may not be a problem, but where water is scarce or presently being overexploited, irrigation efficiency improvements may result in lower crop water productivity, or reduce the efficiency of water use at the basin level. For example, during the late 1990s, China pursued a water savings program that included large investments in sprinklers to irrigate low-value crops on the plain. However, in the dry windy climate, evapotranspiration (ET) increased significantly. Now, China has plans to scale back on this program and start directing water savings activities toward integrated actions that result in ET reduction (that is, real water savings). Thus, improvements in irrigation efficiency need to be associated with other measures to improve water productivity to achieve real water savings. This can be accomplished through an integrated set of engineering, agricultural, and management measures as discussed below.[12]

Basin water efficiency can be improved through evapotranspiration management.

Managing irrigation efficiency can improve the availability of water to the plant root, but it may not be the best option for improving overall basin or scheme water-use efficiency. Canal seepage and deep percolation, for example, return water to the hydrologic system by recharging the aquifers. This water becomes available to the environment and to third-party users. The Madhya Ganga project in India, for example, diverts surplus water during the monsoon season through unlined canals, from which seepage recharges groundwater. Farmers have benefited from the corresponding reduction in pumping costs, the saving in capital cost of well deepening, the more reliable water supply during both seasons, and the improved cropping pattern. Average net farm income has increased by 26 percent. In addition, the recharge benefits domestic and industrial water users in the area.[13]

The integrated water resources management approaches discussed in chapter 3 allow irrigation efficiency to be seen within a holistic basinwide efficiency context. For example, in water-stressed basins, classic measures of water productivity relating crop or income to unit of water withdrawn or applied to the plant root should be replaced by "evapotranspiration management approaches." The objective is not reduction in irrigation losses but reduction of losses to the system through ET to reduce the amount of water removed from the water balance. ET management uses the same range of techniques as conventional approaches to maximize returns to ET through soil and water management, crop management, and so forth. This approach has been successfully piloted in China in the Tarim basin (see box 4.2) and in the Water Conservation Project, and is now planned to be applied in the highly stressed Hai basin.

Crop yields per unit of water can be increased further through on-farm water, land, and crop management practices.

On-farm water management can greatly increase water productivity. Through irrigation scheduling, the farmer can control the quantity and timing of water delivery to align water application with the most sensitive growing periods, and can manage the moisture in the soil through a variety of techniques. For example, cultural and agronomic practices that reduce water depletion, such as different row spacings and the application of mulches, improve water productivity. Irrigation methods also affect these evaporation losses. Drip irrigation, for example, causes much less soil wetting than sprinkler irrigation. Another field-level method for increasing water productivity is deficit irrigation, where deliberately less water is applied than that required to meet full crop water demand. This is appropriate where water, not land, is the limiting resource, which is normally the case in water-scarce areas. Deficit irrigation should result in a small yield reduction in kg/ha that is less than the concomitant reduction in kg/m^3 of water, creating a gain in water productivity per unit of water depleted. China has shown that reductions of 50 percent in irrigation water in wheat and corn growing areas,

Box 4.2. The China Tarim Basin II Project

The Tarim basin is a river basin under stress in a desert area of northwest China. The objectives of the Tarim Basin II Project were to increase farmers' incomes sustainably while reducing water allocations, and to restore environmental flows to the "green corridor," an area of natural beauty and lakes that had dried up three decades before, because of massive irrigation development upstream.

Satellite imagery gave a clear picture of the pattern of beneficial and nonbeneficial ET in the basin and was used along with other methods to assign reduced water quotas to water user groups and to the riverine environment. The knowledge of ET also allowed the project to identify the most productive investments that would save water to achieve optimal basin-level water efficiency, including engineering, agricultural, and management investments. Canals were selected for lining if their leakage was mainly going to nonbeneficial ET. This was often the case because the leakage was contributing to high water tables and salinity, and water was being lost to capillary flux and ET from the ground surface in areas around the canals. Canals where losses were mainly returning to the river or groundwater systems were not lined. Under the project, geomembrane lining along with concrete was used and nearly zero seepage was achieved.

The differentiation of ET at the crop and farm level allowed the project to draw up land and water management plans and inventory the ET requirement of each crop, and even to give advice to individual farmers on ways to improve ET management.

The water quotas are now strictly enforced. The result has been that farmers' usage has decreased and their incomes have increased by 42 percent. Water deliveries are now made periodically to the green corridor, averaging 350 million cubic meters per year, and this important environmental area is being restored. A local resident said, "At the beginning of the Tarim project, everybody used to fight about water. Now everybody understands they have a role to play, can participate in decision making, and uses their energies in managing their water efficiently in accordance with their reduced quota."

Source: Personal communications from Douglas Olson, World Bank, March 2005; and Geoff Spencer, World Bank, March 2005.

when scientifically applied, can result in yield reductions of only 10 percent. Biomass is reduced significantly, but grain production is kept high. A recent review of 82 literature sources (Zwart and Bastiaanssen 2005) found that deficit irrigation can increase crop water productivity while saving water. Based on findings from as wide a range of countries as China, India,

Niger, Turkey, and the United States, crop water productivity may double where irrigation is reduced from 500 mm to 150 mm for wheat, and from 700 mm to 280 mm for maize. In addition, deficit irrigation could lower production costs. Laser land leveling can also improve water productivity.

Crop management practices can also improve water productivity through selection of appropriate crops and cultivars, particularly crops that yield more per unit of water (see box 4.3), and through weed control and integrated pest management. Higher-yielding crops are being developed mainly through germ-plasm improvements to plant efficiency. Improvements may

- match the growing cycle to availability of water or absence of pests,
- increase the rooting depth to help the effective use of water stored in the soil profile,
- increase drought tolerance,
- enhance photosynthetic efficiency, or
- increase the harvest index by increasing the usable portion of the plant's total biomass.

Developing plants with shorter growing cycles can also significantly increase water productivity.

Soil management can also improve water productivity. Techniques include planting methods (on raised beds, for instance), minimum or no tillage, and nutrient management. Levels of fertilizer use in many irrigated systems of Sub-Saharan Africa, for example, are low and there is considerable potential for increased productivity. In Madagascar, for example, minimum recommended fertilizer dosage on irrigated rice is 200 kg/ha, but actual use nationwide averages only 10 kg/ha.

Box 4.3. Developing Less-Water-Intensive Rice Production Systems

Current rice production systems are extremely water-intensive. In fact, 90 percent of agricultural water use in Asia is used for rice production. The International Rice Research Institute (IRRI) estimates that it currently takes 5 m^3 (5 tons) of water to produce one kilogram of rice. Modern rice varieties have about a threefold increase in water productivity compared with traditional varieties. Progress in extending these achievements to other crops has been considerable and will probably accelerate following the recent identification of the underlying genes.

Source: CropLife International 2004.

All factors need to be managed together in an integrated approach to optimize water productivity. Integrated management of soil, water, nutrients, and improved cultivars can raise yields and increase crop water productivity. For example, nutrient management depends on identifying the needed nutrients based on crop needs and soil deficiencies, and scheduling applications at the right time in the growing cycle and in conjunction with water availability. Improved nutrient management can increase water productivity by raising the yield proportionally more than it increases ET (see box 4.4). Combined management of drainage, irrigation, and cultivation practices can greatly improve water productivity in areas affected by waterlogging and salinization. In Megati County in the Tarim basin in China, farmers have been able to quadruple yields and at the same time reduce ET by about 30 percent through a combination of lowering water tables through reduced water application and improved drainage with many of the other factors discussed here (improved cultivars, improved irrigation systems, land leveling, improved fertilization, integrated pest management). The proper management of all these factors can produce more output for less water at the farm level—and at the basin and global level. Key to this approach will be the knowledge transfer agenda—how to transfer technology and techniques to farmers. Translating gains from research to the field through both market approaches and through extension will be a high priority in coming years.

Box 4.4. Fertigation—Fertilization and Irrigation Working Together

Fertilizers raise water-use efficiency by increasing rooting depth and density, as well as the crop's ability to withstand drought stress. However, application is often wasteful. A major breakthrough in targeting the application of crop nutrients came with the development of "fertigation," or feeding crops water-soluble minerals through the irrigation water. Well-nourished plants are better able to absorb the water they need to grow, so fertigation improves the efficiency of irrigation, thus reducing water consumption. In a series of field trials carried out in India and Thailand, crops receiving fertigation produced yields that were between 120 percent and 200 percent higher than those receiving conventional fertilization and irrigation. The technique is so far little used, so its potential positive impact is still largely unrealized.

Source: CropLife International 2004.

Income per unit of water can be increased further.

The choices described above are typically driven by the incentive structure (see chapter 5), and by farmers' abilities to manage risks that result from these choices. Choices begin within the farmer's field with the selection of the crops to grow, with priority to those crops that produce the highest net margin within a manageable level of risk. Farmers can also to a large extent manage their costs and the postharvest handling and on-farm processing stages of production, including postharvest losses, all of which provide the farmer with significant opportunities to add value to his or her production and so increase "net income per drop."

Irrigated cropping intensities can also increase.

A relatively low-cost and environmentally undemanding way to expand the irrigated area is through double (or even triple) cropping. For all developing countries, irrigated cropping intensities already average 127 percent, with China and other East Asian countries averaging 150 percent and more. Potential to drive cropping intensities higher does still exist: FAO estimates that world averages could reach over 140 percent by 2030 (FAO 2003d). Cropping intensity in East Asia as a whole could reach almost 170 percent (figure 4.5). These increases will depend on both supply and demand side factors. They require water availability and therefore are not practical in water-scarce areas, because they give rise to extra withdrawals, but do not usually require extra infrastructure. They will also require widespread

Figure 4.5. Irrigated Cropping Intensities, 1997–9 and 2030

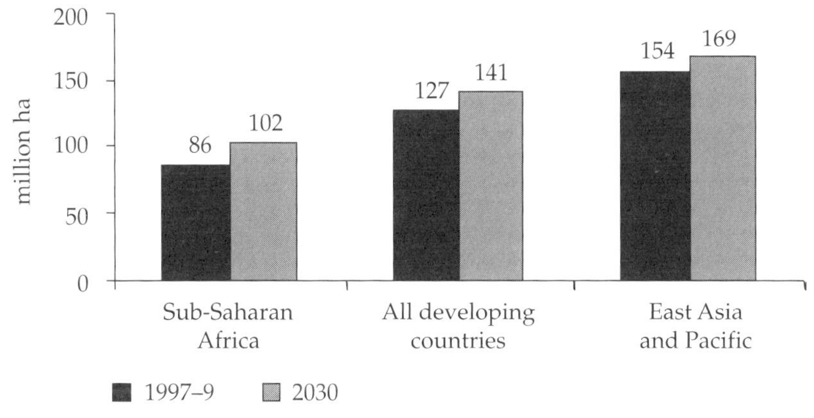

Source: FAO 2003d.

system modernization, which underlines the importance of both the institutional and investment agenda for large scale irrigation and conjunctive use. On the demand side, market incentives will be key, as will be the knowledge agenda, because intensified cropping requires new short-cycle varieties, water management techniques like conjunctive use, and other methods new to most farmers.

4.3 THERE IS CONSIDERABLE SCOPE FOR IMPROVED WATER MANAGEMENT FOR RAINFED AGRICULTURE.

Rainfed farming is expected to produce almost half of increased agricultural production in developing countries in coming years, and improved water management is essential. At present, rainfed agriculture accounts for 60 percent of agricultural output in developing countries. In the FAO projections, 43 percent of the increment in production (between 1997–9 and 2030) in the developing countries will come from rainfed agriculture (see table 4.5). Under the IFPRI "business as usual" projection scenario, rainfed farming in developing countries will have to provide 30 percent of increases in global production (see figure 4.6). The message of both studies is clear—improvement of production in rainfed areas, including through better water management, will be an increasing focus.

The water productivity challenge in rainfed farming is how to introduce accessible technical solutions to improve AWM without increasing risks. Rainfed systems in developing countries tend to be characterized by low productivity caused by low (and variable) water availability, and environmental and soil problems of salinity, temperature, and lack of nutrients. The technological solutions available are characteristically low-yielding: the innovations of the Green Revolution depended largely on water availability, and offered little to marginal rainfed areas. In addition, rainfed sys-

Table 4.5. Share of Rainfed and Irrigated Production in Total Crop Production in Developing Countries
(percent)

Indicator	Arable land		Production		Cereals production	
	Rainfed	Irrigated	Rainfed	Irrigated	Rainfed	Irrigated
Share in 1997	79	21	60	40	41	59
Share in 2030	78	22	53	47	34	64
Share in increment 1997–9 to 2030	67	33	43	57	27	73

Source: FAO 2003d.

Figure 4.6. Share of Irrigated and Rainfed in Cereal Production Increase, 1995–2025

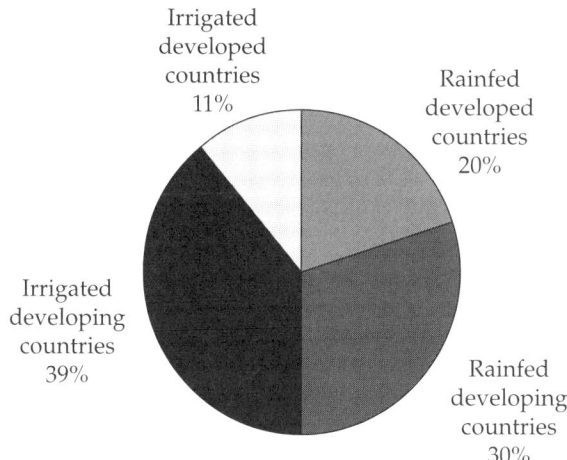

Source: Rosegrant, Cai, and Cline 2002b.

tems are highly vulnerable to risks, including climatic and hydrological risk (drought, floods, climate change) in addition to market risk and land- and water-tenure risk.

There is scope for improvement of water management and soil moisture conservation in rainfed farming. There are known techniques such as soil moisture conservation techniques (minimum and no-till systems, manuring, mulching, recycling city and household waste, and so forth). Water harvesting—collecting water in structures ranging from small furrows, terracing, and bunds, to dams—allows the farmer to conserve rainwater and direct it to crops. Water harvesting can boost yields two to three times over conventional rainfed agriculture. Introducing improved varieties and better cropping patterns, and using minimum tillage methods that conserve water may further increase yields. In mountainous areas, terracing is an efficient farming practice for water harvesting and moisture conservation to grow annual and perennial crops and vines. Evidence shows that local investments in rainfed agriculture can help farmers conserve soil moisture by extending the time water remains inside the productive system and by maintaining or improving organic matter content of the soil.

Some technologies for rainfed areas can have high returns (World Bank 2005b). Investing in supplemental irrigation—a "just-in-time" dose of water—can have a significant impact on rainfed systems. Returns to water in supplemental irrigation are higher than in conventional irrigation.

Farmers readily adopt the techniques once they are convinced they are profitable and reduce risk. Simple low-cost technology is available: in Africa and Asia, simple treadle pumps costing only US$50–$100 can irrigate up to 0.5 ha using family labor. Combined soil and water management investments in rainfed systems can also have a high return. The Loess plateau watershed rehabilitation project in the Yellow River basin of China demonstrated on an area of 1.5 million ha that profitable rainfed farming could be compatible with soil and water conservation. In many cases, technology is available but is not adapted to local agro-economic and sociological contexts. Participatory and bottom-up approaches (see box 4.5) are important to both adaptation and adoption of technology in marginal and rainfed areas.

4.4 LIMITED EXPANSION OF THE IRRIGATED AREA CAN TAKE PLACE.

The potential for expansion of the irrigated area is limited but does exist in a number of countries. FAO (2003d) has estimated that to meet rising demand, the irrigated area in developing countries needs to increase by 40 million ha between 1997–9 and 2030, an increase of 20 percent. Increases will be moderated by water resource availability, by the existence of suitable sites, and by economics. Suitable sites do exist in many countries. For developing countries as a whole, the irrigated area still only covers half the estimated potential (see chapter 1). In Sub-Saharan Africa, in particular, where there is technically ample scope for expansion of irrigation, only

---

Box 4.5. Participation Aids Innovation in Rainfed Systems

Recent pilot projects in India have successfully tested integrated soil, water, and agronomic investments in marginal watersheds. Earlier top-down attempts to introduce new technical packages—for example, vetiver grass—had failed. Instead, new investments were based on a "bottom-up" approach, testing and evaluating innovations. Innovations were first piloted, then scaled up. Cost sharing cemented ownership. Uptake has been excellent. Family incomes increased considerably. The projects demonstrated a cost-effective investment mechanism for making a large and sustainable impact on the lives of poor people.

Source: World Bank 2005b.

Sudan and Madagascar have considerable irrigated area already developed. Ethiopia, by contrast, has enormous water resources potential and very low levels of water infrastructure: artificial reservoir storage is only 43 m³ per person compared to 750 m³ in South Africa—and 6,150 m³ in North America (see figure 4.7).

The high cost of large-scale infrastructure in Africa may make some investments unviable, but as markets develop, smallholder and community irrigation is likely to become increasingly important. By contrast, the countries of the Middle East and North Africa, South Asia, and East Asia and the

Figure 4.7. Water Resources Infrastructure in Ethiopia

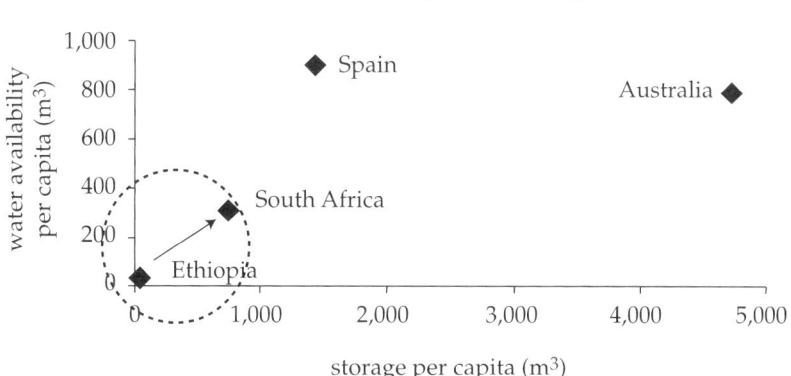

Water availability versus storage

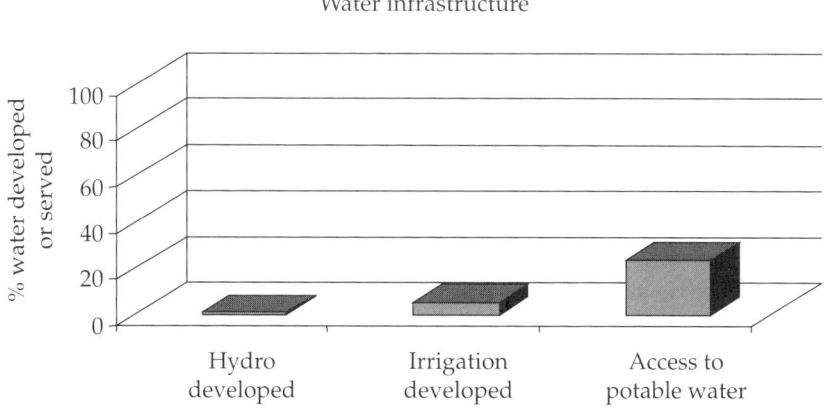

Water infrastructure

Source: World Bank forthcoming.

Pacific have developed two-thirds of their irrigation potential. In Asia, the Mekong basin has some scope for expansion, but other basins, especially in China, are moving toward the limits of potential for expansion. Some increase in the irrigated area can, in fact, be supplied by diversion from structures already in place. In Morocco, for example, existing dams have the capacity to irrigate about 160,000 additional ha but the downstream irrigation infrastructures are not yet in place. In Iran, there are many dams where the downstream investments in irrigation are not yet constructed.

Economic constraints will be increasingly important. As countries and basins approach the limits of their irrigation capabilities, the remaining development lands are likely to be high cost, and will raise difficult economic and environmental problems such as groundwater depletion, drainage of wetlands, or diminution of environmental flows below critical points (FAO 2003a).

Irrigation expansion may occur mainly in Asia—but is most needed in Sub-Saharan Africa. FAO (2003d) hazarded to project in which regions the additional 40 million ha might be developed. Surprisingly, most of the increase is projected to occur in India (13 million ha, 32 percent of the total increase) and China (8 million ha, 20 percent of the total increase). Sub-Saharan Africa and Latin America, despite their potential, are not projected to expand their irrigated areas very quickly. FAO bases the counterintuitively low projection for Sub-Saharan Africa on low population densities, abundance of land for rainfed extensification in some countries, and high cost and low returns to large-scale irrigation development. However, the abundant water resources in countries such as Ethiopia may trigger more interest in irrigation development—and economic returns may be much higher than thought when all off-site benefits are factored in (see chapter 6). Elsewhere in Sub-Saharan Africa, the chronic and deteriorating poverty situation should drive more investment in AWM, both large-scale and smallholder oriented. Extensification has helped to keep Sub-Saharan Africa poor: improved water management and investment in irrigation could help put the continent back on an agricultural growth path to economic takeoff.

4.5 RESULTING INCREASES IN WATER WITHDRAWALS FOR IRRIGATION MAY STRAIN THE WATER RESOURCE BASE.

As indicated in chapters 1 and 3, irrigated agriculture in developing countries already accounts for 40 percent of all crop production and almost 60 percent of cereals production. FAO predicts that up to 2030, irrigated agriculture in developing countries will account for over 70 percent of the projected increase in cereals production.[14] These increases in irrigated production will require large quantities of low-cost water: FAO forecasts that demand for irrigation water will increase by 14 percent by 2030 (from

2,128 billion cubic meters [Bcm] to 2,420 Bcm). Part of the increase in demand will come from higher cropping intensity and part from expansion of the irrigated area (FAO 2003d) (figure 4.8). The International Water Management Institute (IWMI) has projected an 18 percent increase in demand between 1995 and 2025. The International Food Policy Research Institute (IFPRI) has comparable estimates: in its "business as usual" scenario, demand for irrigation water is projected to increase by 12 percent between 1995 and 2025 (Rosegrant, Cai, and Cline 2002b). However, IFPRI projects that actual consumption may increase by only 4 percent due to supply constraints.

In the future, as extra water is withdrawn for irrigation, growing water scarcity will have to be managed. Globally, increased water withdrawals for irrigation by 2030 would make hardly any dent in renewable resources. If the FAO estimate of 14 percent more water needed is correct, the percentage withdrawn would increase from 7 percent to 8 percent. However, for water-short countries and basins, any increase in withdrawals will intensify pressures on the resource, affecting both surface and ground water. That pressure will be localized in basins and aquifers where withdrawals are nearing the limit of the physically and economically exploitable resource. For example, in the Near East and North Africa, irrigation withdrawal would increase from 53 percent to 58 percent of the renewable resource and in South Asia from 36 percent to 41 percent (table 4.6).

Figure 4.8. Projected Irrigated Land Expansion by Region, 1997–9 to 2030

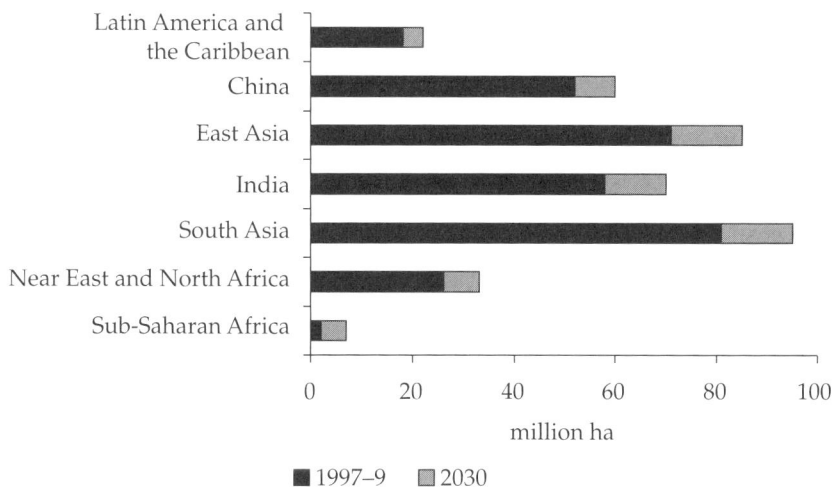

million ha

■ 1997–9   □ 2030

Source: FAO 2003d.

Table 4.6. Annual Renewable Water Resources and Irrigation Water Requirements in Developing Countries, 1997–9 to 2030

Indicator	Sub-Saharan Africa	Near East and North Africa	South Asia	East Asia	Latin America and the Caribbean	All developing countries
Precipitation (mm)	880	181	1,093	1,252	1,534	1,043
RWR (Bcm)	3,450	541	2,469	8,609	13,409	28,477
Irrigation water withdrawal						
1997–9 (Bcm)	80	287	895	684	182	2,128
as % of RWR	2	53	36	8	1	7
Irrigation water withdrawal						
2030 (Bcm)	115	315	1,021	728	241	2,420
as % of RWR	3	58	41	8	2	8

Source: FAO 2003d.

Note: RWR = renewable water resources. Bcm = billions of cubic meters.

At the same time, demand for water from outside agriculture will increase sharply. Because of rapid population growth and rising per capita water use, total domestic consumption will increase quickly. IFPRI projects an increase in domestic and industrial demand of 71 percent from 1995 to 2025, of which more than nine-tenths will be in developing countries (Rosegrant, Cai, and Cline 2002b). India alone will add 340 million people to its urban population between 1995 and 2025. In addition, there will be a demand for environmental uses. As a result, FAO (2003d) projects that by 2030 one in five developing countries—mostly in the Middle East, North Africa, and South Asia—will be suffering actual or impending water scarcity. Worldwide, 10 countries (including the Arab Republic of Egypt and Pakistan) are likely to be withdrawing more than 40 percent of their available resources. A 40 percent ratio is reckoned to be high water stress and 80 percent is very high water stress. In most parts of the world, the water available to irrigation will be constrained and irrigation consumption will grow much more slowly than consumption in municipal and industrial uses. In Asia overall IFPRI/IWMI (Rosegrant, Cai, and Cline 2002a) projects that water consumption by all users will increase by 14 percent by 2025, but irrigation consumption in developing countries will go up by only 4 percent—and in water-constrained China, irrigation consumption is even projected to decline (table 4.7).

Table 4.7. Projected Increases in World Water Consumption, Total and Irrigation
(billions of cubic meters)

Region	Total water consumption			Irrigation water consumption		
	1995	2025	% increase	1995	2025	% increase
Asia	1,059	1,206	14	920	933	1
China	291	329	11	244	231	− 5
India	353	396	11	321	332	1
Latin America and the Caribbean	131	170	13	88	97	11
Sub-Saharan Africa	62	93	50	50	63	13
West Asia and North Africa	135	162	12	122	137	12
All developing countries	**1,358**	**1,603**	**12**	**1,164**	**1,216**	**4**

Source: Rosegrant, Cai, and Cline 2002b.

Water shortages are likely to be a critical problem in a number of important basins. Water in many basins will be inadequate to meet the increase in irrigation demand. In the populous and highly developed basins of South Asia and China, withdrawals as a share of total renewable resources are already very high, in some cases over 100 percent: nonrenewable groundwater is being drawn down (figure 4.9), and nonirrigation demand for water is growing quickly.

As a result, the reliability of irrigation water supply is likely to decrease, particularly in highly stressed basins such as the Haihe basin in China, which is an important wheat and maize producer and serves major metropolitan areas, and in the Ganges basin in India. IFPRI estimates that if nothing changes, several of these major basins will go into permanent groundwater overdraft, threatening water quality and ultimately exhausting the water resource on which they depend (see figure 4.10). By contrast, in the relatively water-abundant basins of Sub-Saharan Africa and Latin America and the Caribbean, water resources will not generally be a constraint to expansion of irrigated production.

The implications of decreasing irrigation water supply availability would be severe for AWM:

• Water stress in irrigation will grow as producers try to intensify production in response to rising demand. There will be a strong push to improve water productivity and to strengthen the use of demand management techniques.

Figure 4.9. Groundwater Withdrawals in Developing Countries, 1995 and 2025

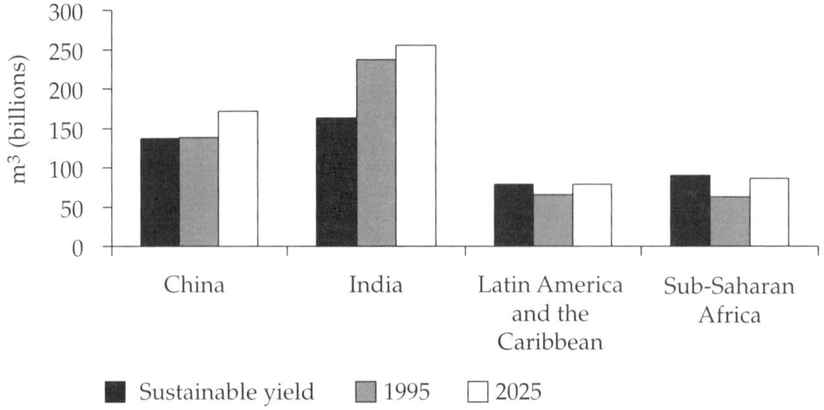

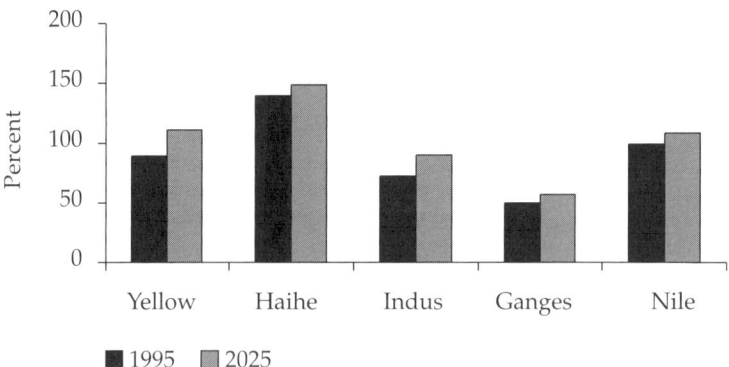

Source: IFPRI 2002.

Figure 4.10. Withdrawals as Percentage of Renewable Resources in Key Basins, 1995 and 2025

Source: Rosegrant, Cai, and Cline 2002b.

- In many basins, intersectoral competition will be intense. Increased with-drawals for irrigation will be limited and mechanisms for allocating water equitably between sectors will be needed.
- Groundwater depletion from increased irrigation will continue and may accelerate. The resource is fully harnessed everywhere except in some

localized aquifers and in some countries in Latin America and Sub-Saharan Africa. By 1995, developing countries as a whole were already in deficit on groundwater, with major deficits in China (25 percent overdraft rate) and India (56 percent overdraft rate). In the North China Plain, groundwater overdraft has reduced aquifer levels by up to 50 meters since 1960. Governments and users will have strong incentives to work together to reduce rates of depletion.

As indicated, some new water withdrawal projects for irrigation would be undertaken. The possibility of harnessing new water resources is highly site specific, and there are a number of water-abundant basins and countries. IFPRI (Rosegrant, Cai, and Cline 2002b) projects that South Asia will harness the most extra water (126 Bcm, 43 percent of the total increase in developing countries). Of this, about 60 percent is expected to be surface water withdrawal, for which major capital projects would be undertaken. Latin America and the Caribbean and Sub-Saharan Africa are also expected to harness more of their underutilized water resources. Careful management will be needed for Sub-Saharan Africa, given the high unit costs and poor implementation record of large-scale irrigation in that region.

4.6 CHANGES IN AGRICULTURAL WATER MANAGEMENT OF THE REQUIRED MAGNITUDE WILL CREATE RISKS FOR THE ENVIRONMENT AND SOCIETY.

Changes in the pace and character of irrigation expansion will create varying but intense environmental risks. As discussed, increased agricultural production over the period to 2030 is expected to come from three sources: one-third from a combination of expansion of arable lands, multiple cropping, and shorter fallow periods; and two-thirds from higher yields. Each of these changes will have a different environmental impact.

Expansion of the irrigated area will largely convert existing rainfed farms to irrigation. Environmental impacts from change of use will need to be managed (box 4.6). Some of the increase may come from drainage of part of the remaining 300 million hectares of wetlands worldwide. In addition, withdrawals from existing surface and groundwater schemes will continue to have an impact on wetlands.

Increases in irrigated area and in cropping intensity will divert more water resources—certainly requiring the construction of additional diversion and reservoir capacity, and increased drawdown of groundwater. Environmental risks will increase, including the risk to environmental flows. The likelihood of continued mining of nonrenewable groundwater was discussed above. Inevitably, if preventive actions are not taken, over-

---

Box 4.6. The Risk of Water Pollution
from Agricultural Sources

Agriculture, especially intensive irrigated agriculture, is the main source of nitrate pollution of groundwater and surface water, as well as the principal source of ammonia pollution. It is a major contributor to the phosphate pollution of waterways and to the release of the greenhouse gases methane and nitrous oxide into the atmosphere. Poorly managed use of pesticides has detrimental effects both on the environment and on human health. Intensification of irrigated agriculture is likely to lead to yet more fertilizer use: FAO estimates that even on conservative assumptions—and if better management practices were introduced—fertilizer use would still increase by 37 percent by 2030.

The risk would come principally from poor management. Nitrogen pollution occurs when application exceeds crop nutrient uptake. In developed countries, the problem is being solved by a combination of technological innovation, regulation, education, and price incentives.

Source: FAO 2003a, 2003d.

---

draft would lead progressively to aquifer exhaustion, saline intrusion, collapse of land structures, and so on.

Environmental flows will come under further threat. In most basins, in-stream flows are still treated as a residual after all other uses have been satisfied, although there have been some significant improvements, at least in thinking and methods (see chapter 3). In some heavily populated basins in China and India, only 5 percent or less of total renewable water is left in-stream. Yet, in-stream flows have important environmental and ecological purposes in the water cycle, quite apart from their amenity value. As water use increases, and as use in some heavily cultivated surface and groundwater basins approaches or even exceeds the renewable resource, environmental flows will continue to be at risk, with the threat of loss of natural habitat, drying up of wetlands, and loss of amenity value.

The risks are likely to grow as intensification patterns follow those in the developed countries. Technical, managerial, and economic tools for managing these risks have been developed in recent years. Their application will be essential as quickly growing demand for food production accelerates the pace of intensification. This will require vision, political commitment, institutional change, and the allocation of financial resources (FAO 2003d; World Bank 2003c).

5
Policy and Institutional Options to Promote Agricultural Water Management's Contribution to Development

This report has outlined the challenges facing agricultural water management (AWM) in a rapidly evolving world, and has indicated the need for changing the way of doing business on a broad front: changes in policies, in incentive structures, in the roles of stakeholders, and in investment patterns. The two final chapters of the report discuss options open to decision makers in these areas. Chapter 5 focuses on polices and institutions, and chapter 6 on investment. The options discussed need to be adapted to each country and local situation and applied in an integrated way. For example, price incentives for water conservation need to be matched with market development policies that allow farmers to invest and to increase incomes. New institutional arrangements for large-scale irrigation management need to be matched with investment programs to improve water service delivery. Managerial and technical improvements at the scheme level need to be matched with sectoral programs that stimulate demand and growth, and these improvements need to be matched with national policies on sustainable and equitable resource management and fiscal prudence, and with global opportunities in trade and technology development.

5.1 GLOBAL AND REGIONAL POLICIES FOR AGRICULTURAL WATER MANAGEMENT

This section reviews the policy and institutional options on issues of trans-boundary water, on the global and regional research agenda, on climate change and hydrological variability, and on the implications of global trade policy and virtual water for agricultural water management.

International riparian issues affect irrigation as the major user of water.

Many countries depend on trans-boundary water flows from the world's 260 major international rivers for large parts of their irrigation water resources (box 5.1).

Box 5.1. Hot Spots for Irrigation and Riparian Issues

The Nile Basin serves Burundi, the Democratic Republic of Congo, the Arab Republic of Egypt, Eritrea, Ethiopia, Kenya, Rwanda, Sudan, Tanzania, and Uganda. As of 1997, the population of this area was 299 million, and is expected to grow 66 percent by 2025. By a 1959 treaty, Egypt is entitled to 55.5 billion cubic meters (Bcm), and Sudan to 18.5 Bcm. These countries have committed almost all these resources to large-scale irrigation. Egypt is expanding its irrigated area by 1 million hectares over the next 20 years, which could require a further 8 Bcm. Ethiopia, where 86 percent of flow originates, has developed only one-twentieth of its potentially irrigable land, and now has ambitious plans for investment in irrigation. Also, hydropower potential on the Blue Nile is enormous. Current technical and political processes are aimed at finding a mutually beneficial solution before pressures become too great.

The Tigris-Euphrates Basin serves Iraq, Syria, and Turkey, covering a population of 103 million (as of 1997), which is expected to grow 51 percent by 2025. Turkey's GAP project (GAP is the Turkish acronym for the Southeastern Anatolia Project) involves more than 20 dams, will irrigate 1.7 million hectare (ha), create jobs for 3.5 million people, and generate vast hydropower. However, GAP could reduce the Euphrates flow into Syria by 35 percent. Syria and Iraq have agreed on water sharing 42:58—but there is no agreement with Turkey.

Source: Postel 1999.

Among countries where irrigation is important, the Arab Republic of Egypt, Iraq, Syria, Turkmenistan, and Uzbekistan depend on rivers flowing through neighboring countries for two-thirds or more of their total surface water. While a basic premise of water resources management is that river basins are best managed and developed as an integrated whole, this always complex task is more difficult in the case of international waters, because there is no "apex authority" through which differences can be resolved and, although criteria for allocating water and benefits can be drawn from a growing body of customary international law, there is no consensus on the criteria for equitable allocation. Nations often seek to develop river segments within their own territories, settling for what are—from an unconstrained, basinwide perspective—second-best investments (World Bank forthcoming). In extreme cases, tensions over international rivers have halted development, as with the Al Wahda Dam on the Yarmouk River in the Jordan Valley, or undermined the viability of infrastructure, as with the Khodaferin Dam, which is under construction by the Islamic

Republic of Iran across the Aras river where the land on the other bank is disputed between Azerbaijan and Armenia. Stresses are rising as water scarcity presses, and perceptions that historic patterns of sharing are inequitable create tensions, especially when there is no water-sharing treaty involving all riparians (Postel 1999). In many areas of possible dispute, the population is growing very fast, and in some, major irrigation programs are underway. These factors will increase both water demand and the scope for tension. The 1997 UN convention on nonnavigational uses of international watercourses established two key principles: (a) equitable and reasonable use; and (b) an obligation to cause no significant harm to neighbors. However, several countries with significant trans-boundary water issues have not accepted the convention—Turkey and China, for example. Agreement on international water sharing can help future development of irrigation in a number of countries—in the Nile Basin, for example. The absence of agreement could compromise these developments.

*Policy and institutional options.*Despite tensions, there is a history of equitable settlement of irrigation and riparian issues through negotiation and joint beneficial development. With support from "honest brokers" such as the World Bank, some best-practice lessons are emerging on enhancing water resources availability through international cooperation on riparian issues, including for groundwater (the Disi aquifer between Jordan and Saudi Arabia, for example). It is essential to act early before the situation becomes critical, and to take advantage of windows of opportunity. In the process, mediation can be key. Solutions need to focus on needs rather than rights, and to create benefits for all—and these need not be water benefits only. Irrigation is a key component of agreements, bringing evident economic benefit, usually as a part of multipurpose operations. For example, under a 1996 agreement around the Aral Sea, the Kyrgyz Republic stores Syr Darya winter flow for release in the spring when Uzbekistan and Kazakhstan need irrigation water. The deal reduces Kyrgyzstan's winter hydroelectric output, so Uzbekistan gives natural gas in return, and Kazakhstan gives coal. The international agreement was brokered by USAID (Postel 1999). Finally, in negotiating assignment of rights, it is important to factor in environmental flows.

Research and technology transfer are vital to obtaining productivity improvements in AWM.

As discussed in chapter 3, the international research system centered on the Consultative Group on International Agricultural Research (CGIAR) plays a vital role in developing research on AWM as an international public good. This role is becoming more important with the rapid advances in science and the widening technological divide between the developed and the developing world. Current technical research programs focus on improving water productivity through new crop varieties; improved cropping systems and on-farm agronomic practices; and better soil, water, and nutrient management. In addition, International Water Management Institute (IWMI) programs are covering key challenges for AWM, which were identified in chapters 1–4: integrated land and water resource management; sustainable groundwater management; water resources institutions and policies; and water, health, and the environment. In addition to this global public research effort, the private sector is conducting extensive research in AWM. Private agricultural input companies are spending more than 20 times what CGIAR spends on research (World Bank 2005a). However, private research is overwhelmingly directed toward profitable, high-tech systems for developed country agriculture.

Policy and institutional options. The "easy" advances of the Green Revolution are giving way to smaller, incremental increases across a broad range of productivity drivers, many of which will have their impact through the combination and integration of a number of technical, institutional, and economic improvements. Further research should cover a broad range of options for improving water productivity. (See table 5.1, which lists just a selection of priority research areas.)

Technical research priorities in the coming years should also include research into salt-tolerant crops that can be grown with brackish water, or with drainage water. A second subject to emphasize is research on water-thrifty crops (see chapter 4), where the following are areas of possible further progress:

- *Increasing the harvest index.* Increasing the proportion of a plant's total biomass that is harvestable yield was the major contribution of the Green Revolution, which raised the harvest index of wheat and rice to around 50 percent. Many plant breeders see little scope for further increase in wheat and rice, but there is surely potential for other crops.
- *Bioengineering.* Bioengineering can breed in traits to close the plant stomata more "promptly" to reduce evapotranspiration.
- *Genetic engineering.* Through molecular marking, scientists should be able to identify the genes that hold desirable traits such as drought resistance and salt tolerance.

Research institutions should forge partnerships with the private sector, which is extremely active in development of irrigation technology. The challenge will be to bring private research down to the level of the needs of smallholders in developing countries and to get affordable technology

Table 5.1. Selected Elements for a Research Agenda in AWM

Category	Option or measure
Technical	Large-scale irrigation modernization and management
	Water management–friendly design of surface irrigation systems
	Low-cost water harvesting technology
	Low-cost water conservation methods in rainfed agriculture
	Low-cost pumping and pressurized irrigation systems
	Surge irrigation to improve water distribution
	Efficient sprinklers to apply water more uniformly
	Low-energy, precision application sprinklers to cut evaporation and wind-drift losses
	Membrane-covered canals
	Adapting drip irrigation to smallholder conditions
	Recycling drainage and tailwater
	Assessment and optimization of irrigation multifunctionality
Managerial	Irrigation scheduling
	Canal management to ensure timely delivery
	Timing applications to suit crop water need
	Water-conserving tillage and field preparation methods
	Land and water management
Institutional	Irrigation management transfer and water user associations for better management
	Irrigation agency reforms
	Water pricing for conservation and efficiency
	Water rights and markets
	Technology development and transfer
Agronomic	Selecting crop varieties with high yield per unit of transpired water (high harvest index)
	Intercropping to maximize use of soil moisture
	Better matching crops to climate conditions and water quality
	Crop rotation respecting soil and water characteristics
	Drought- and salt-tolerant crops
	Crop breeding: water-efficient varieties

Source: Postel 1999; Authors.

to market. Adaptive research of techniques should be further developed, preferably in partnership with manufacturers and suppliers seeking mass markets for irrigation equipment or planting materials (box 5.2). Experience shows that commercial technology can be downscaled in this way but it needs a process of adaptation and market development in which development agents like nongovernmental organizations (NGOs) can play a catalytic role.

Box 5.2. Affordable Drip Irrigation

Microirrigation was developed in the 1960s and has spread rapidly, to cover an estimated 2.8 million ha worldwide—a fiftyfold increase over the last 20 years but still only about 1 percent of the world's irrigated area. The potential for expansion is enormous: in India the current area under drip is less than 200,000 ha, but the potential is in excess of 10 million ha. On-farm water productivity generally doubles with drip. Studies in India show drip generally cuts water use by 30–60 percent and boosts yields by 5–50 percent. However, high cost has been a barrier to widespread adoption.

An international NGO, International Development Enterprises, developed a drip system that costs only a quarter as much as conventional systems by making drip lines portable, and so allowing each line to serve 10 rows of crops instead of just one; by replacing expensive and sensitive emitters with simple holes punched in the drip line; and by using off-the-shelf plastic containers and cloth filters. The resulting system costs only US$250/ha. Field tests on vegetables in the hill areas of Nepal and on mulberries in Andhra Pradesh showed that the system doubled the area under cultivation with the same volume of water.

Source: Postel 1999.

Factoring climate change and hydrological variability effects into AWM

Hydrological variability can undermine growth and poverty reduction strategies. The dramatic challenge of climate change to AWM was discussed in detail in chapter 3. Much of the long-term effectiveness of investment in poverty eradication and sustainable development is vulnerable to this threat. However, AWM programs can be a valuable adaptation option in countries affected by adverse effects of climate change on water resources and by extreme climate variability. In Ethiopia, for example, the priority for investing in AWM is higher when climate-change impacts and hydrological variability are factored in. Indeed, here extreme hydrological variability is echoed in the economic performance. The vast majority of Ethiopia's population (80 percent) subsists on rainfed agriculture, linking the welfare and economic productivity of the majority of Ethiopians to the volatile rains. The correlation between rainfall and overall GDP is strong, as can be seen in figure 5.1.

Based on these insights, a model was constructed to quantify the economy-wide impacts of Ethiopia's water resources endowment, variability, and management under different assumptions of rainfall variability. The model

Figure 5.1. Rainfall Variation and GDP Growth

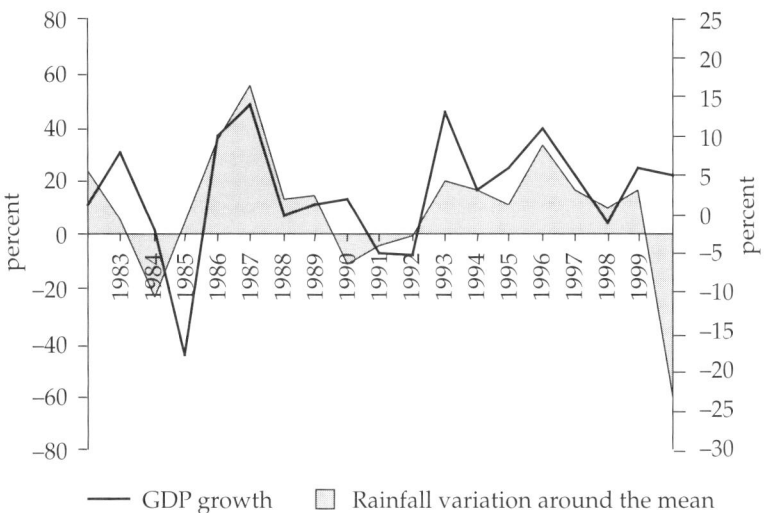

Source: World Bank, SIMA and African Rainfall & Temperature Evaluation System data.

shows that hydrological variability costs the Ethiopian economy over one-third of its growth potential and causes a 25 percent increase in poverty rates, clearly demonstrating the extraordinary impact of drought, particularly rainfall variability, on the Ethiopian economy. Yet, risk management policies and investments can mitigate impacts. Including hydrological variability in modeling and project planning doubles returns to irrigation and drainage, and when irrigation investment is combined with other rural infrastructure investments, GDP growth rates in Ethiopia have been projected to double, from 1.75 percent to almost 3.5 percent (see figure 5.2), bringing a 12 percent decrease in poverty rates (World Bank forthcoming).

Policy and institutional options. The key message is that rising risks and uncertainties should be dealt with by a risk management approach. At the strategy and policy level, climate change risks and hydrological variability need to be built in to economy-wide modeling and to project planning. Adaptation to climate change also needs to be factored into poverty-reduction strategies. Climate risks should routinely be assessed in country and agricultural water sector work, alongside other risk assessments, as was done for the Ethiopia Country Water Resources Assistance Strategy (World Bank forthcoming).

At the field level, monitoring and assessment systems need to be set up. Several donor agencies, including the US Agency for International

Figure 5.2. GDP Growth in Ethiopia under Conditions of Variable Rainfall

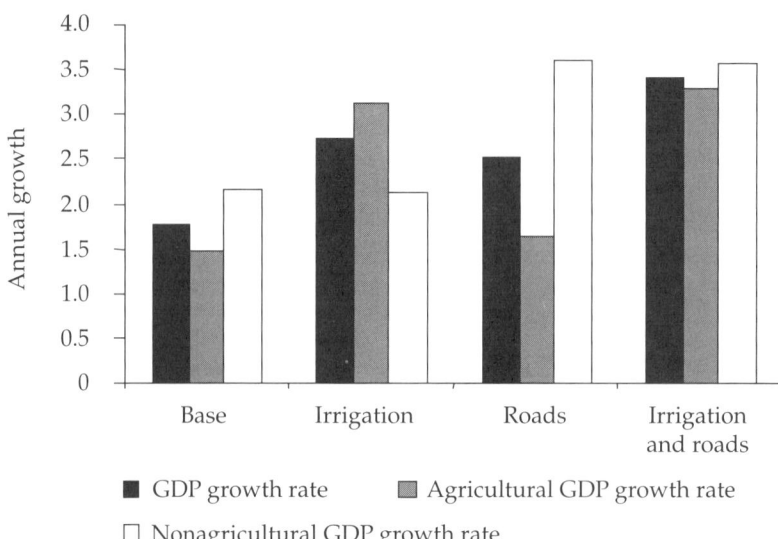

■ GDP growth rate   ▨ Agricultural GDP growth rate

□ Nonagricultural GDP growth rate

Source: World Bank forthcoming.

Development (USAID) and the World Bank, are developing assessment tools and approaches to deal with climate change impacts, especially in AWM. These approaches assess the vulnerability of the area and of the specific subsector to climate change, and identify adaptation measures to include in project design. In particular, the World Bank is developing a rapid climate change risk assessment tool, designed to carry out a preliminary screening of projects. This will include a Climate Risk Management Knowledge Base (possibly extended to include all natural hazards). For projects showing a "red flag," further risk assessments would be performed, and possible adaptation measures carefully analyzed.

USAID's Famine Early Warning Systems Network (FEWS-NET) has helped countries address issues of drought and food insecurity. FEWS-NET analyzes remote sensing data and ground-based meteorological, crop, and rangeland observations to track the progress of the rainy seasons in semiarid regions to identify early indications of potential famine. It also works to strengthen capacity, inform decision makers, and develop policy-relevant information in the regions of Africa and Afghanistan where it is operating. The FEWS-NET Web site (www.fews.net) serves as a gateway of information about threats and updates on response measures.

AWM can also play a part in improving the situation, by stabilizing greenhouse gas concentrations in the atmosphere. As discussed in chapter 1, irrigated agriculture contributes to greenhouse gas emissions. However, it also represents an opportunity for mitigation. Increasing crop yields to limit the spatial expansion of agriculture, more efficient use of energy-requiring inputs such as fertilizers and irrigation, and minimum tillage and tree-based production systems are all ways to reduce greenhouse gas emissions. Most of these actions are win-win solutions, because they help to reduce emissions while increasing profits for farmers.

Possible investment support systems include strengthening agronomic research for drought-tolerant crops, increasing awareness and training in risk assessment, and promoting packages to reduce risk through changing husbandry practices and planting material. Early warning systems may tell farmers when a reduction in stream flow or a drought is expected. Investments in water storage infrastructure and systems (including small storage tanks for rainwater, artificial aquifer recharge, and the like) and water-saving irrigation systems (drip irrigation, for instance) can reduce risk, and these investments have increasing rates of return under conditions of hydrological variability.

Global trade policy and virtual water for AWM

Chapter 3 described the constraints to irrigated agriculture in developing countries that stem from a world trade order still characterized by high levels of developed country agricultural subsidy and widespread restrictions on market access. As noted in chapter 3, progressive reduction of tariff barriers and improved market access could bring substantial economic gains to developing countries, especially with a parallel reduction in domestic trade-distorting support (World Bank 2005a). The present section examines likely impacts of freer trade regimes on irrigated agriculture in developing countries, and reviews policy options.

Trade reform policies will strongly influence water productivity in agriculture. As water for agriculture becomes scarce, more investment would typically be allocated to increasing supply. The marginal cost of supply would rise, reflecting both more expensive technologies and the rising scarcity or opportunity cost. There would be direct impacts on the cost of living and effects on the environment, also. Trade opens up the possibility of reallocating scarce water to higher-value uses, for example, from import-substituting cereals to fresh fruit and vegetables for export. By aligning the prices of inputs and outputs with border prices, removing restraints on international trade will strongly and directly influence water productivity, through the effect on the prices of inputs and outputs. Open trade thus, in principle, improves the returns to water and promotes market-based agriculture that then creates wealth and jobs.

Trade can also create "global" water savings, as the virtual water concept demonstrates. As water scarcity grows worldwide, the opportunity cost of water will rise in many countries, and the "savings" in water that can be obtained by trade in "virtual water" embedded in agricultural produce will become more apparent. Virtual water trade is importation by water-scarce nations of their least water-efficient crops from countries that have a lower opportunity cost of water and higher productivity. Virtual water trade generates water savings for importing countries, and generates global real water savings because of the differential in water productivity between the producing and the exporting countries (Oyebande 2004).

Growing wheat in India, for example, uses four times more water than in France (figure 5.3). In another example, transporting 1 kg of maize from France to Egypt transforms an amount of water of about 0.6 m^3 into 1.12 m^3, which represents globally a real water saving of 0.52 m^3 per kilogram traded. In 2000, the maize imports of Egypt and the related virtual water transfer thus generated a global water saving of about 2.7 Bcm, about 5 percent of Egypt's total water consumption. Globally, the trade in virtual water is rising rapidly, from 450 Bcm in 1961 to 1,340 Bcm in 2000, 26 percent of the total water required for food. Global water savings are estimated at a net 385 Bcm.

Globally, the trade in virtual water is rising rapidly. IFPRI/IWMI calculates that projected increased developing-country cereals imports between

Figure 5.3. The Amount of Water Used to Grow Food
(liters of water evapotranspirated per kg of food)

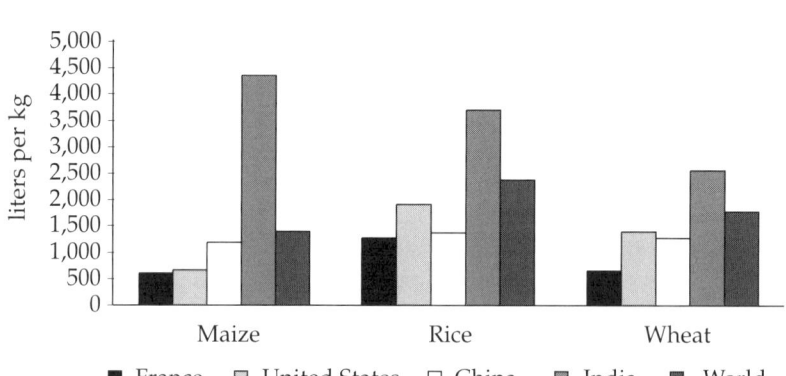

Source: Oyebankde 2004.

1995 and 2025 will save 147 Bcm of water, equivalent to 12 percent of their irrigation water consumption (Rosegrant, Cai, and Cline 2002b).

Policy and institutional options. The impacts of trade reform on irrigated agriculture should be carefully assessed before reforms are undertaken. Gains from trade liberalization are demonstrated by numerous empirical studies. The increase in aggregate welfare of developing countries from global agricultural trade reform could be US$142 billion annually (World Bank 2005a). However, free trade has complex impacts on the entire economy. These impacts can be negative, including for irrigated agriculture. In Morocco, for example, one study showed that while the nation as a whole would benefit from agricultural trade liberalization, those benefits would be concentrated on the urban population; farmers—particularly poor farmers—stood to lose (see box 5.3).

Trade liberalization has important implications for water management choices, too, because water will be allocated to those crops that pay the best returns. Where water is scarce but priced at well below its opportunity cost, for example, liberalization could produce the counterproductive outcome of reallocation of water to less-water-efficient crops. The specific impacts on irrigated agriculture need to be assessed by a modeling exercise. Independent technical assistance to help developing countries assess these impacts is advisable given the complexity of the issues involved (Roe et al 2004; Tsur and Dinar 2004).

Typically, economic mechanisms and social support programs should be developed to help the adjustment toward free trade. Irrigated agriculture will require special consideration, as in the Morocco case, if the transition from protected to free market agriculture is to be accomplished with maximum economic benefit and minimum social cost. Mexico is managing the transition to free trade under the North American Free Trade Agreement by replacing producer price subsidies with subsidies that encourage efficient agriculture and promote social protection (box 5.4).

Although support, in principle, should be required most in nonirrigated areas where farmers have fewer alternatives, the predominance of cereals in irrigated areas (more than half the irrigated area in both Morocco and Mexico—see chapter 4) means that irrigated farmers will also require support, because price competition will be exceptionally fierce for these commodities. The composition of the support package needs to be designed through a technical process matching instruments to objectives. Stakeholder participation, as in Mexico, strengthens design and acceptance.

Environmental impacts also require consideration, because market prices do not reflect the social cost of water depletion, pollution, and other detrimental by-products. A particular problem has been groundwater depletion as new horticulture export markets have opened up.

Box 5.3. Morocco—The Dilemma of an Irrigated Agriculture Constrained by Lack of Market Opportunity

Morocco is at present loosely bound by long-term World Trade Organization deprotection commitments. A more immediate dilemma is presented by the negotiation of bilateral trade agreements: in one, with the European Union (EU), Morocco is preparing to open up part of its cereals market to EU cereals exports, in return for a quota for higher-value horticultural exports to the EU. The agreement would intensify Morocco's long-time move toward a high-value, irrigated export agriculture. The benefits of this trade reform seem considerable: a more efficient allocation of resources, as labor and capital are reallocated to the nonfarm sector; lower food prices; water reallocated to higher-value crops; and agricultural exports predicted to rise by 39 percent.

Yet, the government has decided to protect the agricultural sector from the effects of the new agreement by maintaining tariffs on cereals imports (it will "sterilize" the price impact of the EU cereals by auctioning them). An economic modeling exercise to gauge the impact of the reforms showed that, while water in irrigated areas would be better used, rainfed lands under low-value cereals would go out of production. The increase in farm incomes from the switch to higher-value crops would be outweighed by the losses from the reduction in lower-value crops. The increase in higher-value crops would be constrained by the ceiling set by the quota with the EU. The model shows that small farmers incur the largest losses.

The government has adopted the "sterilization approach" while considering the implications of administering such a shock to its agriculture sector, which is home to most poor people and where standards of living have not improved for years. Like Mexico, which was faced with a similar dilemma under the North American Free Trade Agreement, Morocco is considering introducing a broad range of trade-neutral support programs for its agriculture sector, including direct income support.

Source: Roe et al 2004; Tsur and Dinar 2004; Authors.

Policy reforms should support the development of high-potential irrigated production activities, such as horticulture. As discussed in chapter 3, horticultural products enjoy an especially advantageous position in world markets and can drive investment, modernization, and technological change in developing country irrigation sectors. Horticultural markets are characterized by two very different technical support requirements. The first is to help developing countries negotiate access for these products to high-value markets like the EU. The second is to help irrigated producers and

---

Box 5.4. Using Temporary Subsidies to
Improve AWM—Mexico

Organisation of Economic Co-operation and Development (OECD) agri-cultural subsidies directly affect the economics of irrigated farming in developing countries by depressing farm prices below market-clearing level, particularly wheat, rice, sugar, and cotton. This in turn reinforces the pleas of farmers in developing countries for subsidies for water, ener-gy, and other inputs.

In Mexico, next door to one of the most heavily subsidized agricultures in the world, the government acknowledged that subsidies were required if Mexican agriculture was to survive competition from US production. The new proposed subsidy structure would promote efficient use of resources, especially of groundwater. After careful documentation of the existing incentive structure, the government and farmers agreed on a "subsidy neu-tral transformation," from a package of perverse subsidies (of fertilizers, pesticides, and water) to a package of virtuous subsidies (for example, investment in water-efficient technology and protection of land quality).

Source: Authors.

---

downstream operators respond to market opportunities. Assistance may be needed with regulations on grades, standards, and safety requirements, and possibly for piloting models that integrate the supply chain from field to supermarket, as developed in countries such as Kenya and Jordan. Complementary programs may be required in market development and in regulation of groundwater extraction (discussed below), as well as tech-nological and investment support to smallholders (see chapter 6).

Nations should invest in institutions and technology so that irrigated agriculture can adjust to market opportunities from free trade. Over time, sig-nals from global markets will drive change and modernization in large-scale irrigation systems to accommodate diversified and evolving cropping pat-terns. This will require investment and management flexibility, and is likely to strengthen the case for the involvement of users in planning, investment, and management of irrigation schemes. Because trade-driven growth will be based on knowledge-intensive agriculture, an agenda will be created for innovation and investment in technology development and transfer.

5.2 NATIONAL POLICIES FOR AGRICULTURAL WATER MANAGEMENT

As discussed throughout this report, water resources for agriculture will be increasingly constrained, and intersectoral competition for water will rise.

At the same time, policy makers are likely to have multiple objectives, not just increased agricultural production. They are likely to give more priority to social equity and environmental protection, while maintaining existing food security and rural development goals. Although the policies will be different in each country and basin, in general it is likely that agriculture will be challenged to increase efficiency so that food output and rural incomes will grow; to surrender water—or at least forgo extra—in favor of domestic, industrial, and environmental needs; and to meet the challenges in a way compatible with good management of the environment and in accordance with social values. This section discusses approaches to meeting these challenges: integrated water resource management approaches and the basin-management framework as an institutional process to guide water resource allocation for agriculture; recovering control over groundwater; and water rights and water markets as mechanisms for improving productivity in irrigation.

Irrigation must be treated within an integrated water resource management framework.

As discussed in chapter 3, with growing water scarcity, the interdependence of irrigation with other water uses is clear. Social interdependence is evident: diverse users have claims on a scarce resource, and society has to balance divergent views and goals of stakeholders and decide on trade-offs. Ecological interdependence of the resource and the wider environment is clear from the complex interactions of irrigation, environmental flows, drainage, third-party impacts, and so on. Agriculture covers between one-third and one-half of the surface area of most basins. Agricultural water use and river basin management are inextricably linked.

Policy and institutional options. Integrated water resource management approaches should be adopted, considering all uses of water within a basin, with a view to maximizing welfare. The basin approach looks at the synergy between different sources and uses and at the dynamic of demand in all the various using sectors, especially the agricultural sector, which is by far the largest consumer of water. A basin plan therefore must promote increasing efficiency in water use, provide for equitable transfer of water to higher-value uses, and protect the environment. Many stakeholders and values are involved and as intersectoral competition increases, emphasis must be maintained on the multifunctional aspects of agricultural water. Agricultural water is not simply a production input, but the key to the preservation of ecosystems, maintenance of the rural economy and way of life, and a prime instrument of poverty reduction. The basin approach must also consider externalities such as the effect of agriculture on other users, human health, and the environment—and vice versa.

The principal focus for irrigation development must be on efficiency and productivity rather than new withdrawals. Working at the basin level can bring important advantages to irrigation efficiency—for example, improved irrigation scheduling to account for rainfall variability, or conjunctive management of various sources, including water of poorer quality, where appropriate. When water is very scarce and demand from other high-value users is pressing, mechanisms for intersectoral transfer have to be in place. These may be either administrative transfers, for example, within water master plans, or market-based transfers through rights-based water market systems (see below). Where new irrigation withdrawals are considered, there must be a framework for reviewing the proposal—and alternatives—within the broad resource-management and socioeconomic framework. A special focus has to be maintained on environmental uses of water, which otherwise are treated as a residual, and on drainage (see chapter 6).

The basin approach is most important in very-water-scarce basins, where demand is growing and where the impact of any changes will be high. Some of the biggest basins in the world are already experiencing water scarcity (Rosegrant, Cai, and Cline 2002b):

- The Yellow River basin supports 136 million people, 11 percent of China's population. The basin contains 13 percent of the total cultivated area of China but has only 3 percent of the water resources. Increased scarcity is shown by interruptions in the flow, declining groundwater levels, disappearing lakes, and silting up of river beds.
- The Haihe River basin supports 90 million people. The basin covers 3.3 percent of China's total area, supports 10 percent of the population, and produces 10 percent of China's agricultural output. Of its 10.8 million ha, 7.1 million are irrigated. The basin has been running a water deficit for 25 years.
- The Indus basin covers 10 percent of the area of India, and 5 percent of the cropland. Water tables have been dropping, and groundwater basins have run dry for parts of the year. Water scarcity is an international issue here: India and China almost went to war after Partition, but an apparently durable treaty was signed in 1960.
- The Ganges basin covers 26 percent of India's area and 30 percent of the cropland, of which 20 percent is irrigated. The Ganges often experiences severe seasonal water stress.

The objective is to achieve basin efficiency. Basin efficiency is water productivity assessed at the river-basin scale, taking account of hydrological flows and reflows basinwide. At the basin level, the measure of productivity is no longer "crop per drop," as at the level of the scheme or the farm, but the ratio of beneficial water consumption in irrigation to total irrigation water consumption at the river-basin scale (Rosegrant, Cai, and Cline 2002b). This allows irrigation water productivity to be compared to the productivity of other alternative uses, and assesses productivity improvements within the context of their effect on other users. The basin approach allows water and land use planning for all uses within the basin.

Irrigation should be managed within a basin planning framework that ensures that water flows to the uses on which society places the highest value. A basin plan should include appropriate policies for management of surface- and groundwater resources, for drainage, and for environmental flows and impacts. It should factor in a policy for nonconventional resources, such as reuse of wastewater, and spell out under what conditions new storage and diversion would take place. It should reconcile competing demands from different users, and specify how other policy objectives, such as poverty and food security objectives, are to be taken into account.

Pathways toward more sustainable agricultural groundwater management

The rapid development of groundwater irrigation and the reasons for its spread were discussed in chapter 2, as were the emerging problems of over-exploitation and depletion. Because the resource is unseen and largely privately developed, governments have until recently tended to ignore the need for its regulation and management. Recovering control is now high on the agenda for many countries. There are three principal reasons for the overdraft problem: competitive pumping from an open-access resource; the lack of an established institutional basis for management; and government support policies that often price energy very cheaply. Attempted solutions to date have tried to deal with each of these facets.

Regulatory approaches have been applied to try to establish user rights and an institutional basis for management. Regulatory and rights-based approaches define access, register and regulate rights, and may even regulate trade in groundwater. Clearly, regulatory and rights-based approaches would be useful, but attempts to implement these approaches have run into significant problems. Groundwater systems are often poorly evaluated and monitored, and the quantitative basis for defining rights tends to be weak. In some countries, the number of wells that would need to be monitored is extremely large, many being located remotely on private land. Water rights systems also tend to be socially complex and often based on deeply embedded cultural values (World Bank 2003a), so that third-party moderation is hard. Finally, enforcement capacity is weak (exemplified by the fact that many wells have for years been illegally connected to the electricity grid). Even if regulatory capacity existed, transactions costs would be extremely high, and regulation would be difficult to implement rapidly over any significant area.

An alternative has been to rely on self-regulation—decentralized collective management of groundwater resources by water users. There are some examples of self-regulation from developed countries, in particular the American West and Spain, where groundwater users have, with varying degrees of success, federated to safeguard the sustainable supply of water (van Steenbergen 2002). Can self-regulation work in developing countries? Developing countries tend to have a much larger number of groundwater users in a weaker governance environment. However, developing countries do often have traditions of cooperation on water management, and may even have workable rules relating to groundwater regulation. There are many examples of local groundwater management, particularly in areas with shallow, semiconfined aquifers. These examples fall into two main classes: norms regulating demand through social pressure; and local regulation by structured water user organizations, which may recognize rights and undertake collective investments (box 5.5). There is, however, a limit to what norms can achieve. Where investment in supply-side infrastructure is required, for example, an organizational structure for aquifer management may be needed. Also, norms and social pressure are unlikely to work where there is stress and some users risk losing out, as in the case of many deep aquifers.

Policy and institutional options. The first best solution for agricultural groundwater management is a rights and regulation framework. A governance system is needed that establishes clear and measurable entitlements and allows self-management by user groups supported by government in resource assessment, regulation, and dispute resolution. This can be implemented in a decentralized way and can be complemented by self-regulation. Experience shows that a workable groundwater management system has intensive user involvement and user-government partnerships.

Rights and regulation approaches need to be supported by an incentive framework because national food and energy policies can exert an overriding influence on the behavior of groundwater abstractors. Among these policies, *energy subsidies* are probably the biggest driver of groundwater depletion, but subsidies on well drilling, pump sets, and grain prices are also significant. Governments need, in principle, to eliminate energy subsidies for water pumping, and redirect those subsidies into water-saving technology or poverty-reduction programs (World Bank 2003a). Clearly, there are massive constraints to such a move, which could have repercussions throughout the economy where energy is fungible and where ostensible policy objectives of poverty reduction or food security have long since been

Box 5.5. Local Groundwater Management

Use of Social Norms in Baluchistan, Pakistan
In the Panjgur Valley in Baluchistan, the irrigation system brought water through tunnels (called *karezes*) from the subsurface flow of the river and from the infiltrated runoff from the surrounding hills. Farmers had seen how rapid development of wells in neighboring valleys had drained the karezes. They imposed a total ban on the development of dug wells and tube wells. However, nobody is excluded: those wishing to develop groundwater may construct a kareze. The limitations on the development of dug wells are easy to understand—typically no well may be dug within 5 km of an existing kareze. After some dispute, drinking water supply wells were exempted from the ban. Implementation is highly informal—each kareze owner has the moral right to intimidate each potential investor in a dug well. If this has no effect, the local administration is approached. Officials invariably respect the rules set up by the community. These groundwater rules are not supported by a special organization, and no attempt has been made to define them individually. The rule is a simple norm—an embargo on a certain groundwater abstraction technology.

A Structured Groundwater Association in Egypt
In Salheia in the East Delta, recharge of groundwater was limited, and well yields and well reliability went down. Seawater started to intrude. In 1993, one farmer organized a get-together of the 400-odd landowners in the area of 1,000 ha. The meeting decided on a hydrogeological survey of the area to determine safe yields and establish a common management system. Following the hydrogeological survey, the farmers decided to continue pumping only from a limited number of wells and to develop a common network of pipelines. The investment in the network was some US$300 per ha, which was to be recouped from water charges. The individual system was thus transformed into a collective asset. The agreement between the farmers led to the establishment of the Omar Ben al Khattab Water Users Association. The association also decided on a ban on new wells in the area. The Salheia case thus moved beyond coordinated individual responses to groundwater problems and even "communalized" groundwater by linking all lands to a common pipeline network. A local groundwater association opens up a large range of management options that do not exist in a social norms–based mode of groundwater management.

Source: van Steenbergen 2002.

eclipsed by the political economy of vested interests. Elements in solving this dilemma include classic political economy moves (political will and leadership, transparency, participatory approaches, and so forth—see below on ways of "managing" the political economy of reform for agricultural water). In addition, economic and technical instruments can help, including step tariffs to charge large users more, support for irrigation efficiency investments, and others. Finally, increasing energy prices can only be a solution if used as part of an integrated package of groundwater management and rural and agricultural development, to sustain the rural economy as pumping is reduced.

A combination of demand- and supply-side measures should always be used. Demand-side investments to improve the efficiency of water use are essential in all at-risk aquifers (box 5.6) and water conservation technology should always be encouraged. As aquifers are depleted, conjunctive use (see chapter 4) will become a frequent option. Complementary local supply-side measures, such as aquifer recharge enhancement, rainwater harvesting, and urban wastewater reuse, are also needed. These investments provide incentives to groundwater users and can provide an initial focus for their participation in aquifer management (World Bank 2003a).

Reducing the decline in aquifer water levels requires savings to be translated into permanent reductions in well abstraction rights and actual pumping, not into increases in the irrigated area. This will require a flexible system of abstraction rights and clear incentives for users to act in the collective interest of resource conservation. If the objective is to transfer water from agriculture to higher-value urban uses, the municipality—or private investor—might finance improvements in agricultural irrigation (generating real water savings) in return for abstraction rights over a proportion of groundwater saved (World Bank 2003a).

Strengthening local water resource management is an alternative. In most countries, governance is relatively weak, groundwater use is too widespread, and the problem is too urgent to await the patient development of the institutional structure for a rights and regulation system. In these cases, strengthening local water resource management may be required, especially in areas with confined aquifers. Social solidarity may help, but the motive can be simple profit, as in the Egyptian example in box 5.5. Organizational structures such as water user associations (WUAs) have a capacity for implementing supply-side measures, but norms-based approaches can effectively restrict groundwater use on a broad scale quite quickly and are much easier and faster to establish (van Steenbergen 2002).

All policy options for groundwater management need to be accompanied by an information, monitoring, and evaluation component, because groundwater management is only possible if knowledge about the resource is available. This is a key area in which governments should invest—even decentralized setups will rely on governments to provide hydrogeological information. Monitoring can be participatory, which makes it a dynamic part of the management process (box 5.7).

Box 5.6. Saving Groundwater through Demand Management

Only those modifications to irrigation and cropping practices that reduce nonbeneficial evapotranspiration or nonbeneficial discharges to sinks or saline water bodies actually represent real water savings (although these components may not be easy to quantify accurately). Thus, the primary aim of agricultural demand management for groundwater resource conservation should be to reduce (a) evaporation from the irrigation water distribution system; (b) soil evaporation from between crop rows; (c) evapotranspiration by the crop itself, ineffective in producing yield; (d) direct phreatic evapotranspiration by unwanted vegetation; and (e) direct evaporation during spray irrigation.

There is generally considerable scope for these types of agricultural water savings through

- engineering measures, such as irrigation water distribution through low-pressure pipes (instead of earth canals) and irrigation water application by drip and micro-sprinkler technology;
- management measures to improve irrigation water scheduling and soil moisture management; and
- agronomic measures, such as deep plowing, straw and plastic mulching, and the use of improved strains and seeds and drought-resistant agents.

If larger water savings are needed, consideration should also be given to changes in crop type and land use (for example, through higher-value crops under greenhouse cultivation, or returning a proportion of the area to dry land cultivation of drought-resistant crops). An even more radical option would be to place a ban on the cultivation of certain types of irrigated crops in critical groundwater areas.

Source: World Bank 2003a.

If overpumping is reduced, significant risks for production and incomes will have to be managed. If highly stressed groundwater regions stopped overpumping and returned to sustainable water use, water availability for agriculture would decline. The International Food Policy Research Institute (IFPRI) estimates that if groundwater worldwide is to be returned to sustainable extraction levels, global groundwater pumping would decline from the 1995 level of 817 Bcm to 753 Bcm, and the total area planted to cereals worldwide would be 730,000 ha less (Rosegrant, Cai, and Cline 2002b). Most of the change in irrigated area would occur in developing

Box 5.7. Accelerating Local Regulation through Participatory Hydrological Monitoring

The Participatory Hydrological Monitoring (PHM) program developed in Andhra Pradesh, India, under the APWELL project provides essential hydrogeological information through participatory monitoring. Farmers are being trained in measuring groundwater parameters themselves. They are provided with

- a drum and a stopwatch to measure the discharge of a number of their wells;
- a water table recorder to measure the depth of the water table;
- a rain gauge, installed in a sheltered place;
- ready reckoner tables and training in how to make crude water balances.

A field hydrologist helps to analyze the results. PHM has helped farmers to adopt more water-efficient crops and practices. Floriculture, castor seed, cotton, and maize have replaced rice, which now accounts for less than 5 percent of the area. Farmers have also been taking steps to improve recharge close to their wells through sink pits and small check dams. The next step is to turn awareness into local resource planning. A government observation well in each village will be monitored by the community. Also, in the last annual government "mass contact" campaign, senior government staff discussed simplified water balances in village meetings. PHM is successfully bringing groundwater knowledge to groundwater users.

Source: Govardhan 2003; van Steenbergen 2002.

countries, especially in China. Production and incomes would also fall, if nothing else changes. IFPRI estimates that total cereals production could decline by an annual average of 18 million tons in the period 2021–5. Crop prices under this "low groundwater scenario" might be 5–10 percent higher in 2021–5 than under "business as usual" projections (Rosegrant, Cai, and Cline 2002b). The world would not go hungry, because developed countries would increase production, but the developing world as a whole would increase net imports, with major increases concentrated in China and India, and there would be significant income losses among the farming population. There is thus a major trade-off to be faced. Plainly, a program to get farmers to reduce overpumping of groundwater has to be accompanied by a significant research and technology transfer effort to get "more from less." This effort would also need to promote a switch to higher-value crops, investments, and policy reforms to increase basin efficiency, and promote programs for "modernization" to move rural people away from water-based activities. All these changes are possible, and recent experience with improving productivity in the Tarim basin of China appears to confirm technical potential for increasing production per unit of groundwater consumed (see chapter 4). However, the key will be to introduce *workable institutional structures of incentives and regulations,* an area where there has been little track record of success, even in developed countries.

Agricultural water management and water rights and markets

Establishing secure water entitlements would greatly help good AWM. Seen from the irrigation farmer's perspective, secure water entitlement is primary. Planning for a year's crop, or planting trees for the future, investing in irrigation equipment, or accessing credit facilities—all depend on water security. Lack of secure water entitlement leaves the farmer open to risks that he or she is unable to manage.

Unclear water rights result in conflict, resource degradation, disincentives for investment, and disproportionate negative impacts on the poor, who rely on communal or open-access resources. Administrative allocation is often inefficient, and a confusing web of administered prices and subsidies obscures incentives and the true opportunity cost of water (World Bank 2005a). Defining rights would increase allocative efficiency and promote water-conserving technologies. In economic terms, the ownership of water rights increases the perceived value of the resource to the level of its opportunity cost. The resulting security of water tenure increases incentives for investment and water conservation (box 5.8).

However, property rights over irrigation water are extremely difficult to establish. The attribution of legal rights on large-scale surface schemes is hard when water quantities are uncertain and difficult to measure, and service delivery weak. The allocation of rights to groundwater is harder still, because it requires technical quantification and monitoring of the resource, and an institutional setup to regulate it. Traditional water management systems coped with water rights in the past. They have, however, had trouble coping with the very different challenges of tube well groundwater extraction.

If rights can be established, providing an administrative system in which trading can occur gives rights holders economic incentives to use their entitlements efficiently. Water can then "flow" to its highest return use. Markets should thus encourage investment in more efficient irrigation technology, promote water conservation, help reallocate water to highest-value uses, and enhance agricultural incomes. Only a few countries have succeeded in helping efficient water markets to develop. In Chile, for example, which has a legally recognized system for trading water, water markets have pro-

---

Box 5.8. Formalization of Water Rights in Peru

In Peru, irrigation is well organized, with 64 water user organizations along the coast, about half of which are technically and financially autonomous with respect to cost recovery and operation and maintenance. The remaining half are well on their way to achieving the same goal. Now irrigated agriculture is being challenged to become more efficient in the export market by the impending free trade agreement with the United States.

To prepare for this, the government has launched a massive effort to formalize traditional water rights. The objective is to ensure that beneficiaries have legal security on the use of irrigation, thus creating incentives to private investment, to operation and maintenance of schemes, and to water conservation.

Users are being organized in irrigation blocks to facilitate the management of these rights, a water rights registry has been set up, and licenses and permits are being issued. Satellite images are being employed.

The first phase is covering the coastal area, followed by a second phase in the Sierra, the Andean highlands. At the beginning of 2004, out of about 1.5 million agricultural water users countrywide, fewer than 3 percent had registered water rights. By early 2005, the program had already formalized 300,000 water rights in the coastal area and the World Bank–financed Coastal Irrigation Project is assisting in formalizing another 190,000 water rights. The program is expected to be complete by 2008.

Source: Personal communication from Peter Koenig, World Bank, March 2005.

---

duced substantial economic gains from trade, particularly in transferring water between urban and agricultural uses. Some developed countries have water trading systems, as in parts of California and with Murray Irrigation Limited in Australia.

Water markets have worked well in some places for a long time—and new ones (both formal and informal) are emerging, albeit slowly. In most traditional irrigation systems, mechanisms for water trading have long existed, and these systems have to some extent adapted to the changing demands of the modern world. In many countries, informal farm-to-town water conveyance (usually by tanker, sometimes by pipe) exists. In some countries, these informal markets have been successfully formalized. For example, in Jordan in the 1980s, irrigation tube wells were rapidly depleting groundwater reserves, the Ramsar Convention wetland at Azraq had dried up, and intermittent supply in the major towns was being supplemented by unlicensed tankers, further depleting the groundwater stock.

The government worked with farmers in a sequenced program to recover control over the resource and to allow farmers to realize the full opportunity cost of their water. All wells were registered, licensed, and metered; the aquifers were characterized and individual quotas assigned to each well; a monitoring and regulatory program ensured compliance; wells were licensed to sell a certain quantity of water to the tanker trade; and the tankers were licensed and controlled for water quality.

In the future, water stress and structural change in the irrigation sector may drive more active interest in water markets, because their development can help farmers constrained by dwindling water resources or hurt by macroeconomic reforms. As scarcity pushes up the opportunity cost and as nonagricultural demand grows, water markets may be a way of easing the transition out of farming. A recent study demonstrated that farmers in Morocco who may be affected by de-protection of cereals markets can be partly compensated by creation of a water market that allows farmers to realize the shadow price of water (Roe et al 2004; Tsur and Dinar 2004).

Policy and institutional options. Demanding legal, administrative, and managerial requirements make tradable water rights and water markets a long-term prospect in most countries. Formal water markets require a physical conveyance system, with volumetric water measurement, clear water rights, an enabling legal framework, and transparent trading rules. Institutional structures are needed to manage delivery and to provide judicial oversight and dispute resolution. Several preconditions—including a strong link between water and land rights, a prior history of informal water trading, a sound legal system, a system for registering water rights, and good governance—make water trading work. Other success factors include an independent regulatory system to allocate water rights and safeguard essential uses, and a good hydrological information base and titling system.

Establishing water rights is the first step. A number of countries are compiling registers of water rights, as in Peru, and this creates incentives to investment and conservation. The development of tradable rights and markets can follow, if there is a demonstrated need and feasibility. There are two possible tracks. One track is to support the development of a formal rights-based system by developing over time a flexible legal framework of entitlement and transfer, with capacity building. An interesting finding from the Chile case is that informal markets can act as precursors, demonstrating the measures needed for a formal market to work under local conditions. The alternative is to formalize existing informal markets, as in Jordan. In either approach, the problems associated with individualization of water rights could be overcome by recognizing the rights of a group and strengthening the enabling and supporting environment for decentralized and community-based approaches to water rights administration.

5.3 AGRICULTURAL POLICY AND AGRICULTURAL WATER MANAGEMENT

As discussed in chapter 1, AWM is a process of resource management to provide one of the essential inputs to agricultural production. Agricultural policy works both on the demand side—exercising a powerful influence over farm incomes through price and fiscal policy—and on the supply side, driving production patterns through investment and incentives. Thus, AWM and agricultural policy interact closely. In the past, policy makers have tended to focus on supply-side and production policies and have neglected the role of signals derived from free markets and prices in motivating farmers to allocate water resources and investment efficiently and thereby increase incomes. This section describes the policy and institutional implications of two essential areas of interaction between agricultural policy and AWM: links between the development of input and output markets and prices and AWM; and links between national food policy and AWM. Closely related issues of fiscal policy, incentives, agricultural incomes policy, cost recovery, and subsidy are discussed in the following section.

Agricultural policy that allows internal and export markets to develop is key.

Development of agricultural markets can drive investment and productivity in irrigated agriculture. At the household level, market development can help drive irrigation modernization and improve water productivity. It can promote investment, generate growth through diversification and productivity gains, increase and diversify incomes, provide employment, and reduce the cost of food and increase its availability. Market development can promote more efficient and less water-intensive crop management practices and higher-value cropping patterns—fruit, vegetables, flowers.

Incentives created by profitable markets are rapidly reflected in private irrigation, particularly groundwater, which can respond flexibly. Adapting large-scale irrigation to take advantage of market opportunities through modernization programs is a bigger challenge, because large schemes generally have less flexible water delivery capability. At the household level, market development can provide a complement to new AWM technology and so reduce poverty. A case study from Zambia illustrates these effects (box 5.9). At the national level, market and trade reform can enhance both domestic food security and national growth, and poorer countries benefit the most.

Box 5.9. Market Links Drive Smallholder Irrigation Investment and Production in Zambia

Despite abundant land and water resources, Zambian agriculture is poor, with weak markets and rudimentary irrigation techniques. The Zambia Agribusiness Technical Assistance Center (ZATAC) has promoted out-grower schemes directly linked to ready markets through agribusinesses. This strategy offers small growers an opportunity to be partners in the value chain and offers agribusinesses a chance to increase their supply base and benefit from economies of scale without the associated capital investment. ZATAC also provides credit for irrigation equipment. For the first time in the history of Zambia, smallholders now grow irrigated fresh vegetables for markets in Europe, thanks to the alliance between small-holder producers and agribusinesses. The combination of market access and simple irrigation technology has released these farmers from the low-income poverty trap of rainfed agriculture.

Source: World Bank 2005b.

Policy and institutional options. For markets to develop there should be, first and foremost, a conducive framework. As discussed in the Agricultural Directions in Development report (World Bank 2005a), domestic market reforms—liberalization, privatization, removing subsidies—are used by many governments as the complement to external trade reforms. The enabling environment should encourage inward and domestic investment and provide for secure contractual arrangements, supported by the necessary legal and financial institutions.

In addition, strategic investment to promote markets can be critical to the development of irrigated agriculture. The recent Ethiopian CWRAS (World Bank forthcoming) highlights the twin pillars of rural poverty reduction in that country—irrigation infrastructure and investment in market development. One such investment is in rural infrastructure, particularly roads. In fact, roads may be among the best economic development and poverty reduction investments that governments can make. Roads can significantly increase competition and reduce costs for both input supply and the marketing of outputs. A second key investment is in research, development, and extension or technology transfer, which need to be carried out in partnership with professional and commercial bodies. A third area is in the proactive development of farmer-market links, a difficult area for governments because this is essentially a market-driven private activity, but one where business-oriented NGOs have some comparative advantage, as in the Zambia case (box 5.9) and in the work of the Central Asia Development Group to promote the revival of horticultural markets in Afghanistan. Significant improvements in farmers' livelihoods can be achieved where the development of irrigation is integrated with investment in roads and markets (box 5.10).

Government's role should go beyond the framework to ensuring that markets and market infrastructure really work. Governments are increasingly taking an active role in developing the "behind-the-border" agenda in trade facilitation in areas ranging from institutional and regulatory reform to improving customs and port efficiency. Government's role is best undertaken in partnership with the private sector so that investments are driven by market demand, and risks and responsibilities are shared. This can be supported by trade facilitation programs. These programs, many of which are undertaken in partnership with the private sector, are aimed in large part at improving quality and timeliness and reducing transactions costs for high-value produce, and so are powerful drivers of efficient irrigated agriculture. Governments should also work with industry bodies on the design of standards, for example, for packaging, and on trade promotion activities.

Box 5.10. Nigeria—National Fadama Development Project

In the early 1990s, Nigeria developed small-scale irrigation through the extraction of shallow groundwater in the *fadama* lands (plains and low-lying areas underlined by shallow aquifers and found along Nigeria's river systems). In addition to 50,000 shallow tube wells, the program invested in 825 km of roads and 126 storage and marketing facilities. Farmers organized themselves into user associations for irrigation management, cost recovery, and access to credit, marketing, and other services. The combination of improved access to decentralized groundwater technology and to markets resulted in significant increases in incomes. Returns per ha in Jigawa for vegetable production increased by two-thirds, and by three times for wheat. For rice paddies in Niger Province, the returns per ha increased fivefold.

The improvements in livelihoods were palpable—and were widely distributed. About 90 percent of farmers in Gomber increased their incomes, while 30 percent enrolled their children in school and 17 percent changed the roofs of their houses from thatch to zinc. Family nutrition also improved.

Source: World Bank 2000b.

Changes in national food policy can drive intensification of irrigated production.

The above discussion of trade focused on the impacts of the global trade environment for irrigation. The present section reviews the interaction between national food security policies and irrigated production.

Food security is likely to become a growing preoccupation for the poorest countries. Developing countries will import more of their food needs in the coming years. According to IFPRI forecasts, while food production will increase much faster in developing countries than in developed countries, it will not keep pace with demand, and food imports will need to increase. The Food and Agriculture Organization (FAO) estimates that agricultural production in developing countries can cover only about 80 percent of the increased food demand in these countries to 2030. The shortfall would cause a widening gap: for the developing world as a whole, self-sufficiency is expected to decline from 91 percent to 86 percent. As a result, the food trade balance is expected to turn sharply negative (US$50 billion annually by 2030, figure 5.4). This will not, in principle, be a problem for higher income countries with rapidly growing nonagricultural economies that will allow the import of cheap food. The poorest countries, however, are unlikely to have the resources to import food but they are likely to promote domestic food production.

"Hot spots" for food trade gaps in the future (according to IFPRI) are Sub-Saharan Africa, where cereal imports are projected to more than triple by 2025 to 35 million tons, West Asia and the Middle East and North Africa, where cereal imports are projected to increase from 38 million tons in 1995 to 83 million tons in 2025. This may be all right for the Middle East and North Africa, because the reliance on water-saving cereal imports in West Asia and North Africa makes economic and environmental sense, provided that it is supported by faster nonagricultural growth. However, this is a problem for Sub-Saharan Africa. It is highly unlikely that Sub-Saharan Africa could finance the projected level of imports internally; instead international financial or food aid would be required. Failure to finance these imports would further increase food insecurity and pressure on water resources in this region (World Bank 2003b; Rosegrant, Cai, and Cline 2002b).

The desire of developing countries to achieve food self-sufficiency has been one of the biggest drivers of irrigation policy in the past, but this may change in the future. The implications for irrigation of the forecasts described above are far-reaching. Self-sufficiency may no longer drive irrigation development in many countries. Where self-sufficiency is no longer possible, reliance on market mechanisms will be an alternative. This has a powerful impact on irrigation, because food self-sufficiency goals allocate scarce irrigation water to low-value-added cereals rather than to high-value crops,

Figure 5.4. Trade Flows between Developing and Developed Countries

Source: FAO 2003d.

reducing the resources and incentives for higher-value production and limiting scope for poverty reduction and economic takeoff. A switch away from self-sufficiency should have a beneficial effect: developing countries will be able to move up the value chain, for example, to export high-value-added, irrigated products to earn the foreign exchange needed to pay for cereal imports. As income levels rise, they may choose to adopt more market-oriented policies, pushed by limited production possibilities at home and the availability of cheap "virtual water" imports. This is a vital question facing countries that could now afford to import more food, but are struggling to retain their self-sufficiency in basic food products (box 5.11).

Policy options and trade-offs. Food security can be increased most efficiently by channeling water and other scarce resources to the uses in which they are most effective in raising the incomes of the poor, not by specifically targeting food production. In some countries, this strategy will result in increasing food production, in others not. In all cases, however, the emphasis has to be on efficient resource allocation, and on the development of markets to add value to the production of the poor and to ensure that food is available.

Better-off countries should consider moving progressively toward high-value irrigated production. As income levels rise, countries that have limited scope for expanding irrigated food production can afford to import more of their food needs. Quickly industrializing countries under water stress, such as China, should consider progressively releasing water to higher-value crops and to other sectors and importing more basic food

---

Box 5.11. China—Struggling with Food Self-Sufficiency Goals as Water Shortages Grow

Water shortage is probably the single most important problem facing China's agriculture today. The past AWM achievement in China is considerable: 22 percent of the world population fed on 9 percent of the world's arable land and 6 percent of the world's water resources. Irrigation produces 75 percent of national food needs and 90 percent of cash crops.

However, water resources are fully developed in many areas, and overexploited in some. Water consumption in excess of sustainable renewable yields is estimated to be 30 Bcm annually in desperately water-short areas in the northern part of the country. In recent years, the strong growth of urban demand has caused agriculture's share of water withdrawals to drop from 85 percent to 65 percent in water-scarce areas. The volume of water available to agriculture is likely to fall in coming years (a 5 percent drop by 2030). At the same time, China's population is projected to increase from 1.3 billion to peak at 1.6 billion by 2040. Food demand is projected to increase at a faster rate (by two-thirds) because of increasing consumption and changing diets. Two-thirds of China's grain production benefits from enough rain to escape shortages. For the remaining one-third, irrigation is needed and water stress is growing under competitive demand. There is still considerable potential for further gains in water productivity resulting from integrated irrigation technology, agriculture, and management measures (see chapter 4), particularly from better water control, more efficient use of water and fertilizer, and reduced postharvest losses. However, as competition for water grows from other sectors, and as overall water consumption is reduced to meet the sustainability imperative, China will have to consider easing its food self-sufficiency objective if economic growth is not to be constrained. The importing of animal feed grains, for which demand is rising quickly, could be less controversial from a policy perspective than the import of food for the Chinese people.

Sources: Speech of the Chinese water minister at World Bank Water Week, Washington, DC, March 2005; FAO 2004b; personal communication from Douglas Olson, World Bank, 2005.

---

commodities. However, decisions on changes in food policy should be made after careful analysis of impacts on both producers and consumers, and should be accompanied by support and safety net programs where needed.

Increasing irrigated food production will enhance the food security of the poorest countries. In the poorest countries of Sub-Saharan Africa, the low resource base, low capitalization, and low alternative production possibilities combine with high levels of risk to make increasing local food production the most efficient way to improve food security. This will continue to drive the policy and investment agenda for irrigation in Sub-Saharan Africa in coming years and these countries should invest whenever possible in new irrigation infrastructure and in the full range of measures to improve water productivity, both in irrigated and rainfed agriculture. At the same time, irrigation development should not be driven by food self-sufficiency goals if more profitable options are possible. For poorer countries in Sub-Saharan Africa, escaping from the poverty trap requires taking some risks in moving toward a market-driven agriculture. Where market opportunities exist—or can be created—for higher-value agriculture, poor countries should assess the trade-offs involved and wherever possible should promote diversification into more remunerative cash crops. The inevitable risks could be underwritten by external partners, who can support investment in high-value irrigation and AWM and who can also provide a safety net in the form of food aid, if needed.

5.4 FISCAL POLICY, INCENTIVES, AND AGRICULTURAL WATER MANAGEMENT

Driven by food production and rural development objectives, governments worldwide have invested heavily in irrigation infrastructure and in subsequent operation and maintenance, financing about half of the US$30–35 billion invested in agricultural water globally each year. Changes in the ways in which governments do business, described in chapter 3, have begun to affect fiscal policy: in general, governments are starting to invest more prudently, to decentralize, to require more cost sharing, and to use demand management instruments, such as incentive structures, alongside supply management instruments, such as investment. In some countries, Mexico and Turkey, for instance, finance ministries have successfully driven reductions in fiscal transfers. In other major irrigation economies, however, the fiscal burden remains very high. In Egypt, for example, the annual public transfer for water-related services is equivalent to 5 percent of GDP. This section first describes the policy and institutional options for governments in setting the incentive structure for AWM. The discussion reflects the dual role of incentives in public policy—to minimize the fiscal burden and to promote productivity of resource use. It covers in turn irrigation water pricing and cost recovery and then other incentives to water conservation and efficiency. The second part of the section deals with broader questions of public subsidy to AWM and mechanisms for reducing the overall fiscal burden while promoting water productivity and increased incomes. Issues of public investment policy for AWM are discussed in chapter 6.

Incentives for irrigation and AWM

Economic incentives in AWM are signals that affect decisions. They motivate water suppliers and users to manage water in line with two central policy objectives in AWM. *Financial objectives* include ensuring that the cost of efficient water service is paid for, and that investments are financed. *Efficiency objectives* include water productivity in irrigated agriculture, resource conservation, and transfer to higher-value uses. In addition, the incentive structure is very heavily influenced by other government policy decisions —on the trade, fiscal, food security, poverty reduction, and investment regimes, for example—so both politics and planning from the broader economy are brought into shaping the agricultural water incentive structure. Trying to get incentives "right" in these circumstances—to achieve the central financial and efficiency objectives—is extremely hard. This section will deal first with the key question of irrigation service charges and cost recovery, especially for large-scale irrigation. The section will then deal with the broader incentive framework that drives farmer decisions to invest in AWM and to produce irrigated crops efficiently.

Irrigation water pricing and cost recovery. As discussed in chapter 1, management of large-scale irrigation has been plagued by problems of irrigation service charges, both low levels of charges and low levels of collection. In Pakistan and the Philippines, less than half the operation and maintenance costs are recovered; in Bangladesh, rates average less than 10 percent. The Operations Evaluation Department (OED) of the World Bank rated the Bank only marginally effective on user charges issues, with the very low cost recovery rates in India a particular problem (World Bank 2002). In many countries, service charges are set by the government or a government agency and so are a political issue. As a result, governments and management agencies set charges at levels far short of costs and are lax in enforcing recovery. Reasons for this diffidence may be objective policy considerations—food security, poverty reduction, rural development, equity, and so forth—or political expediency. On the user side, low payment is often linked to dissatisfaction with the quality of service and lack of ownership of scheme-level decision making, both of which are frequent results of the predominant government role. In stark contrast, privately managed schemes (about half of the world's irrigated area) have no problem as both capital and recurrent costs are by definition fully recovered.

Low cost recovery leads to poor service. Underfunded scheme managements give poor service, and poor service in turn reduces the economic and financial viability of the scheme, reduces farmer income, and reinforces reluctance to pay. Where cost recovery is low, schemes fall back on the government budget, which is often an unreliable source subject to annual appropriations and unrelated to need or performance. Decentralization and locally accountable management become difficult, because large-scale irrigation management is accountable to its government paymaster. Underfunding of scheme operation and maintenance causes the service to deteriorate over time and the government may also have to invest in rehabilitation.

Policy and institutional options. In designing a system of irrigation service charges, there is first a need to be clear about the objectives. Usually these would be to ensure cost recovery adequate to sustain operations and maintenance (a financial objective). Less frequently, service charges are designed to maintain and sharpen incentives to water productivity (the efficiency objective).

In deciding who pays for water service, all cost recovery programs have to relate to the basic facts—the sustainability of irrigation systems rests on the twin pillars of demand-responsive service and cost recovery adequate to the system needs. If systems are to deliver quality service, somebody has to pay for it. Neither government, nor irrigation agency, nor irrigator has an interest in a system that cannot deliver, and there has to be clarity and commitment about who is to pay for the service. Typically, the irrigator should pay a fair share on the "user pays" principle that individuals who benefit from public investment and scarce resources should pay. The advantage of the irrigator paying is double, because water charges not only bring in revenue but signal opportunity cost. However, if constraints mean that a full cost-recovery policy cannot be introduced, the alternative needs to be made clear. If irrigators cannot pay, then government must.

There has to be a decision on what costs should be recovered from whom, and why. A cost recovery policy should start with an analysis of the full range of services and benefits produced and allocate project costs among all beneficiaries, including those outside the irrigation network who benefit from positive externalities or from nonirrigation services of a multifunctional scheme—environmental, water resource management, hydropower, water supply, landscape—and then assign costs accordingly. Typically, upstream works such as headworks and primary canals have public-good elements that justify government funding, while downstream works at the tertiary and quaternary level deliver entirely private benefits that justify the irrigator paying. For the secondary canal level, sharing of the costs may be appropriate. Consideration should be given to whether to recover all costs—initial capital costs, replacements, and operation and maintenance costs—or only a portion. Typically, recovery might be based on replacement cost and operation and maintenance cost only. The analysis should also look at the ability to pay, which is likely to range from about 5 percent to 30 percent of net farm income, depending on quality of the service.

Higher charges are not anti-poor. One study of 26 irrigation systems in six countries in Asia (ADB/IWMI 2004) found that "reasonable" charges for water did not handicap poor farmers because systems that were financially sustainable worked better with regard to delivering water equitably. The study found that annual charges for irrigation water varied greatly, from under US$5/ha to US$67/ha. Charges were considerably lower in South Asia than in Southeast and East Asia. According to the study, South Asia's low water charges trigger a cycle of poor irrigation performance, leading to low agricultural productivity and the perpetuation of poverty. Fees in this region tend to disappear into central government coffers and are not earmarked for recycling to irrigation managers for investment in improved system performance. Where fees were higher and a decentralized management system was in place, water delivery was considerably more equitable. High rates of fee collection in the systems studied (for example, 95 percent in Indonesia) suggested that water fees are accepted by users (ADB/IWMI 2004).

The choice of charging mechanisms is dictated partly by objectives of scheme financial sustainability and efficient water use and partly by practical considerations of whether the mechanisms can be implemented. If cost recovery is the overriding objective, there is a range of mechanisms. If water is scarce and reducing water use is an objective, a volumetric charge is in principle more appropriate, but only where it is physically possible and not too expensive to implement. In practice, different mechanisms may achieve similar results—and often implementation (that is, maximizing collection) is the key consideration (Easter and Liu 2004). Mechanisms providing for recovery of water charges include:

- *Area basis*, which is simple to set up and administer, and is appropriate when water is not scarce and most farmers are growing crops with similar water requirements. More than three-quarters of the world's large-scale irrigation area uses this system.
- *Volumetric*, which is good for efficiency and easy to understand but costly to implement. It may not recover full costs if, for example, there is a shortage of water or if demand drops.
- *Tiered block system*, which charges a low "lifeline" tariff for a basic quantity, and progressively higher tariffs for extra quantities. This system is appropriate when water is scarce and farm incomes are low, but it is hard to set the blocks at the right level to achieve the multiple policy objectives of full cost recovery, efficient use, and protection of the poor.
- *Two-part tariff*, which recovers fixed costs by a flat rate "admission charge" and recovers variable costs and promotes efficiency by a flat rate payment for all units. It has proved hard to set the rates and hard for farmers to understand the system.

Moving to implementation. Implementing a number of basic irrigation sector reforms helps to greatly improve implementation of cost recovery. The service agency should have the autonomy and financial accountability that creates both reliable service and incentives to collect. The supply agency should have the autonomy to respond to farmer needs, adapting service and delivering water when farmers need it, and in the right amount. The agency should be able to keep revenues it collects and to use them to improve the service, investing, for example, in improved infrastructure to provide better water control (unlike, for instance, India, where revenues go to the Treasury). This has an important effect for farmers, because one of the prime causes of farmer dissatisfaction is the lack of an observable relationship between charges and service. It also has an important effect for the service agencies, because autonomy will give incentives to the agency not only to improve services but to maximize collections, on which its ability to operate and stay in business depends. Autonomous agencies have more flexibility in applying "carrots and sticks" to farmers, and may also set up incentive schemes to encourage their staff to collect charges (Easter and Liu 2004).

For example in Awati, China, an institutional reform established a financially autonomous irrigation district. Now, in Awati, staff salaries are completely dependent on the water charges they collect. Following the reform, the collection rate shot up to 98 percent. In Bayi Irrigation District, China, the staff receives rewards if they can turn in the fees before a set deadline, and are fined if the payment is late (Easter and Liu 2004).

User associations should be involved systematically in all water management decisions, because their participation will improve cost recovery, among other benefits. Where farmers have more authority and responsibility over water management, usually through WUAs, recovery rates have generally improved. In Mexico, collection rates reached more than 90 percent. In Turkey, collection rates improved to 76 percent after irrigation management transfer began. Successful pricing schemes show the benefits of a participatory approach (box 5.12).

Typically, WUAs should provide a mechanism for mutual accountability through which associations can participate in designing the cost recovery system and improving collections, and can also ensure that fees collected are used to maintain and improve the system. In addition, user participation should help reduce costs by improving efficiency of water use, because it provides incentives for responsible management and conservation. Where user participation works well, it has reduced politicization of issues such as irrigation service charges.

The management of services and of cost recovery should be efficient and transparent—a key factor in farmers' willingness to pay their water charges. Fee structures should be equitable, administratively simple, and easily

---

Box 5.12. Using Block Tariffs to Conserve Water and Improve Environmental Quality

Broadview Water District in California wished to reduce the amount of drainage water discharged to the San Joaquin River, but internal reuse of the large volumes of drainage water led to growing salinity in soils and water delivery. The Board devised a two-tier structure that charged farmers a relatively low standard rate for volumes of water up to 90 percent of historical average water use, and a rate two-and-a-half times larger for extra water. The objective was to encourage improved on-farm irrigation practice and to reduce the volume of drainage water. The block tariff was accompanied by a low interest loan program for investments in water management equipment. As a result, irrigation efficiency (crop water requirement less effective rainfall divided by water delivery) improved from 70 percent to 85 percent over 10 years. Farmers reduced water use, total use on the scheme fell by half, and volumes of drainage water dropped. Success factors included a participatory approach that built an incentive structure founded on empirical facts from water use efficiency in the past and which farmers regarded as fair; the accompanying loan program; and simplicity, with just two tariff blocks that the farmers could easily remember and monitor.

Source:" Experience in implementing economic incentives to conserve water and improve environmental quality in the Broadview Water District, California" at www.worldbank.org/grouponei.

---

understood by users and by those administering fee collection. System transparency includes farmers' ability to see how much water they received, how their payments are used, and how water charges are determined. With efficient service and efficient charging combined, collection rates can be very high. Recovery rates for piped groundwater reached 100 percent with the integrated circuit card systems used in China's Shandong province (box 5.13). In systems providing good service, advance payment for water, as in Tunisia, would ensure full recovery of charges. In Sindh, Pakistan, by contrast, farmers are not willing to pay because the financial system is not transparent and they do not see that the charges paid are used to deliver a good service. The farmers said that they were willing to pay for services, but not for "someone's wife's jewelry" (Cornish and Perry 2003; World Bank 2005b IN0104; Easter and Liu 2004).

Efficiency improvements should be introduced to reduce costs and expand the revenue base. Farmers' reluctance to pay charges because of poor irrigation service delivery can be overcome not only by participatory and transparent processes of charging and recovery, but by real reductions in cost and

---

Box 5.13. Automated Irrigation Charge Collection System in Shandong, China

Shangdong is one of the biggest agricultural provinces in north China. Irrigation water accounts for 70–80 percent of the total water use, but water is scarce. Farmers must purchase a prepaid card to operate an automated "integrated circuit" (IC) machine, which measures and controls the groundwater release. After each irrigation, the farmer receives a receipt, stating the amount of water used, the price paid per unit of water, and the total deducted from the IC card. All servers are connected by the internet, so they are easy to control and monitor while the administrative costs are greatly reduced. The cost of each server is 1,000 yuan (US$120), paid for in a single year by the value of the water saved under the new system. With more than 200,000 IC servers provincewide, the province saves about 5 Bcm of water annually.

Highlights of the system's features include the following:

- The IC machine gives farmers full control over water use, which promotes efficiency.
- The system effectively enforces payment collection. The system has achieved 100 percent collection rates.
- The water charge is on a volumetric basis, which encourages reduced water use.
- The system greatly reduced administrative costs because personnel are no longer required to collect fees or open and close gates and the end user is charged directly.
- The amount of water used is accurately recorded, and the charges are transparent.

Source: Easter and Liu 2004.

---

by improvements in farm and scheme financial performance. In the irrigation reforms in Victoria, Australia, for example, some 80 percent of the improvement in financial performance came from efficiency gains and an expanded revenue base, and only 20 percent from increased water prices to irrigators.

Benchmarks and targets for cost recovery in large-scale irrigation should be established through international dialogue. Cost recovery is central to the future development of the irrigation sector, and progress has been limited. The International Commission on Irrigation and Drainage (ICID) recently prepared a proposal for "[f]ive principles for sustainable cost recovery in irrigation," based on conceptual work and on experience of recent reforms (box 5.14). These five principles should be the point of departure for a sus-

Box 5.14. ICID's Five Principles for Sustainable Cost Recovery in Irrigation

Transparency of cost recovery

- Define and clarify the nature of the services provided: for example, upstream and downstream limits, drainage, flood protection, and so on.
- Identify all direct and indirect beneficiaries of the service, so that a fair sharing of the costs may be realized.
- Most importantly, clarify contracts between service providers and users so that service agreements formulate effective accountability mechanisms.

User empowerment

- Identify an effective interface for dialogue by using formal negotiations over an acceptable price, having regular contacts between service provider and beneficiary, creating a clear agenda, and making an adaptable schedule.
- Identify forces that work against equity by protecting the poorest farmers against the most powerful users.

Sustainable cost recovery

- Plan measures to improve cost recovery, even though full cost recovery is unlikely.
- Prioritize the costs to be covered, for example, reimbursement of loans, cost of maintenance and renewal, personnel costs.
- Identify potential emergencies and crises and discuss solutions.

Economic incentives toward best practices

- Meter water.
- Use quotas, rationing, and pricing as incentives to encourage water allocation compliance.

Clear policies

- Separate service provider from regulation authority. Set up an external body to assess the quality of the services provided and to act as a control on government participation.
- Clarify exactly what integrated water management means to the stakeholders so that their responsibilities are clearly defined.
- Separate agricultural policy from water policy. Clarify and delineate the nature of irrigation services because these are often caught between the two policies and conflicts can create barriers to sustainability in irrigated agriculture.

Source: Tardieu and Prefol 2002.

tained international and national effort to define internationally valid benchmarks and best practice in irrigation cost recovery.

Other incentives to water conservation and efficiency

As water scarcity increases, more irrigation projects will have to reduce water use per ha and to invest in water use efficiency. Where water is scarce, farmers in theory have an incentive to use water efficiently. However, where water is cheap or water rights are insecure, farmers will not invest to save water or they may use saved water to expand their farms. In principle, water charges on a large-scale irrigation system can incorporate efficiency incentives. However, few governments are willing to raise water charges to the level required. Also, where alternative crops or technologies are not readily available, water demand will be inelastic, and the price may have to go up significantly before it can affect farmer behavior (box 5.15). Under those circumstances, other instruments and incentives may be considered (Easter and Liu 2004). This section reviews incentives other than price incentives.

Policy and institutional options. Where the demand curve is inelastic and where there is a water shortage that imposes water saving, *rationing* (in the short term) or the allocation of *quotas* (for the long term) should be considered effective ways to reduce demand and encourage efficiency. Quotas work better than prices when water users are not very responsive to water price changes (as in the Iran example). A quota reduces water consumption by creating a high shadow price. In surface irrigation, a quota would be a fixed allocation of water shares to different canals and to water users sharing water from the same canal. In groundwater irrigation, a quota system would specify an annual rate of extraction for each water user (Easter and Liu 2004).

Box 5.15. Nonprice Instruments to Promote Water Use Efficiency

At Zayandeh Rud, Iran, water was increasingly scarce and there were no alternatives to the low-value crops the farmers were growing. Water prices would have had to be raised twentyfold before farmers would invest in field technologies to improve water use efficiency. This rise would have brought water charges to a level equivalent to two-thirds of current farm revenue, putting farmers out of business. One alternative proposed was to use water charges only to cover operation and maintenance costs and to use rationing to restrict water use and encourage investment in water saving.

Source: Easter and Liu 2004.

One result of improved service is to win farmer confidence in the reliability of water service, and so reduce field-level storage behavior and over irrigation. For example, following system reform in Katepurna, India, farmers no longer flood their fields in the dry season, because irrigation scheduling is planned ahead according to water requirements and soil type. Farmers have saved 7.7 million m^3 of water annually and have expanded the irrigated area from 2,027 to 3,646 ha (an 80 percent increase) (Easter and Liu 2004).

Public education campaigns should be considered as a way to make farmers aware of water scarcity and explain to them why water should be treated as an economic commodity. This is especially important in places where people traditionally view water as a free good and a basic right. In many projects, public education programs have been combined effectively with price increases. Using the example of Katepurna, India, again, the principles of irrigators organizing for management and the need for efficient water utilization were promoted through newspaper, radio, exhibitions, pamphlets, and posters. Slogans on participatory irrigation management and efficient water use were written on compound walls, canal structures, offices, and public buildings to promote collective action. To motivate irrigators, cultural groups were formed and cultural programs (songs, drama, and the like) were arranged at village level. This helped improve the community's understanding of the value and importance of irrigation water (Easter and Liu 2004).

Transfer of assets can also act as a powerful incentive. Farmer ownership of assets reduces transactions costs, and increases farmers' willingness to invest (FAO 1999). Other incentives to water conservation and efficiency may come from establishment of water rights and water markets (discussed above) and from "smart subsidies" (see below).

The best approaches are likely to be *packages* that contain both positive and negative incentives. For example, changes in water prices could be linked to allocation of water rights, which could make the package easier to pass and implement. Transfers of assets, management by users, and capacity building form another powerful package. Incentives to use more efficient technology can complement all approaches, although they need to be carefully designed. Jordan used the tariff system both to give incentives to water conservation and to achieve full cost recovery in an equitable fashion, while providing positive incentives to irrigation efficiency through technology transfer and investment subsidies (box 5.16).

Reducing the fiscal burden

Following the discussion above on sharing of costs in large-scale irrigation, the present section reviews the rationale and practice of the whole

---

Box 5.16. Using a Mix of Incentives in the Jordan Valley

Though aware that water use in the large-scale surface irrigation scheme in the Jordan Valley was inefficient, the Jordanian government was reluctant for political reasons to increase prices in the 1990s. An integrated approach was adopted whereby every farmer had a quota of water at a relatively low price, and a step tariff system obliged larger users to pay more. The tariff system was calibrated to cover the costs of operating and maintaining the system. A parallel program provided incentives for more efficient water use through technology transfer and lower priced irrigation improvement equipment. Thus, local physical, economic, and political factors contributed to an integrated incentive package.

Source: Authors.

---

range of subsidies to irrigated agriculture and makes recommendations. As discussed in chapter 1, capital and recurrent subsidies for irrigation have been almost universal. It is estimated that farmers receiving water from government-built irrigation projects seldom pay more than 20 percent of the water's real cost (Sur, Umali-Deininger, and Dinar 2002). One estimate for India is that canal irrigation is subsidized nationwide to the extent of 95 percent if all attributable capital and recurring costs are taken into account. Irrigators with their own wells also receive subsidies on agricultural water, through low prices for diesel fuel, electricity or equipment. Irrigators also receive subsidies on factors of production other than water, and also often receive output price subsidies via protection policies (Rosegrant, Cai, and Cline 2002b; Sur, Umali-Deininger, and Dinar 2002).

There are arguments for limiting subsidies as water grows scarcer. Subsidies are in principle inefficient for AWM, because low-cost water gives little incentive for improving productivity. Each subsidy risks creating distortions in the market such as capital bias or crowding out of market competition. Subsidies are hard to eliminate, and their fiscal cost is often exorbitant—the annual irrigation subsidy in Egypt is US$5.0 billion, but benefits from their reduction can be substantial (box 5.17). Without an exit strategy, the fiscal burden mounts. Subsidies also create negative environmental impacts for which society at large generally has to pay.[15] Table 5.2 shows the harmful impact of irrigation subsidies on the environment. Finally, subsidies are typically "anti-poor," because generally the better off benefit. A study in India estimated that only 13 percent of Indian households had access to canal irrigation, so only that proportion of the population benefited from the extensive subsidies. Among that 13 percent, two-thirds

---

Box 5.17. Reduction and Targeting of Irrigation Subsidies in Haryana, India

The Haryana Water Resources Consolidation Project, appraised in the mid-1990s, was the third major irrigation project in the state. Two previous projects had failed to achieve objectives related to the reform of irrigation service charges.

During project formulation, detailed discussions were undertaken with the government and users, identifying and specifying the total expenditures required for operations and maintenance and replacement; the limited areas where subsidies would continue; the policy for differential charges between sectors; and the link between investment components that the project would finance and the related recovery of costs.

By the end of the project, water services in the state were fully self-funding. This success was dependent on several factors: clear definition of service standards and of priorities for water allocation (formulated as part of a State Water Plan, under the project); a well-managed irrigation system; a participatory approach; and an effective, legally enforced revenue collection system.

Source: World Bank 2005b.

Table 5.2. Environmentally Harmful Consequences of Irrigation Subsidies

Subsidy	Mechanism through which subsidy may harm the environment	How it may harm the environment
Surface water price	Overuse of water and cultivation of water-inefficient crops. Use of inefficient technologies.	Pollution and depletion of water bodies. Salinization, elevated levels of water tables, and drainage problems.
Energy price	Substitution of surface water with groundwater, especially in places where surface water supply is inadequate or irregular.	Overuse of groundwater due to excessive pumping. Groundwater levels are lowered; aquifers are depleted and contaminated via intrusion of low-quality water from adjacent aquifers or seawater intrusion.

Source: Sur, Umali-Deininger, and Dinar 2002.

were "marginal farmers"—but they received only 27 percent of the sub-sidy. The better-off farmers, less than 5 percent of the population, received 73 percent of the subsidy (World Bank 2005b; Sur, Umali-Deininger, and Dinar 2002).

Policy and institutional options. Subsidies should be justified by showing that the benefit from them surpasses the cost. There are legitimate public policy objectives in AWM that cannot be obtained by a market or regulatory approach and that could be obtained by public subsidy. For example, subsidies are often paid at the farm level for goods and services identified as of public interest such as maintaining terraces that contribute to groundwater infil-tration and soil conservation. At the level of communities, measures to enhance landscapes may be subsidized or watershed management investments may be implemented. At the national level, subsidies may be granted for such public interest activities as research and extension in ecologically sensitive AWM. However, certain drawbacks and costs often make subsidies a second best option, including implementation costs, the difficulty of targeting, and the difficulty of avoiding perverse outcomes. In general, these costs and risks outweigh the benefit in AWM: trade reforms combined with reform of eco-nomic incentives in agricultural water use, such as water pricing reforms or promotion of water markets, improve welfare more than subsidies (Roe et al 2004). Justifying subsidies for AWM may thus be difficult except in cases of environmental externalities or income redistribution where even the most intelligent market-based approach cannot achieve the public policy goal.

Where governments decide to continue subsidies in AWM, guidance on design should be drawn from past experience. First, subsidies should be justified and designed within the framework of a coherent incentive policy package, as discussed above. The policy should cover not only sub-sidies through direct incentives such as irrigation service charges but also those transmitted though indirect incentives such as diesel pricing and trade protection policy. Second, where a subsidy is decided on, design and targeting are critical to ensuring that the policy objective is attained and the subsidy goes to the intended beneficiaries. Third, the costs should be cal-culated over the life of the subsidy and an exit strategy should be incor-porated. The current generation of "smart subsidies" now being introduced—for example, cost sharing subsidies on efficiency-improving technology such as drip irrigation—incorporate many of these lessons (Easter and Liu 2004). However, though justified on grounds of encour-aging innovation and compensating for externalities involved in water conservation, even these subsidies can bring their own distortions. Box 5.18 gives an example of crowding out in India, which was ultimately suc-cessfully overcome by market developments that rendered the subsidies redundant.

Box 5.18. Irrigation Efficiency Subsidies Slow Adoption of Drip Technology

Drip and sprinkler technologies have been aggressively promoted in India since the mid-1980s; yet, today, the area using them is only 60,000 ha. A big part of the problem is subsidies that, instead of stimulating the adoption of these technologies, have actually stifled the market. Subsidies have been directed at branded, quality-assured systems, but in the process have not allowed viable, market-based solutions to mature.

Subsidies are channeled through the big irrigation equipment companies. Their equipment typically costs US$1,750/ha, which puts it out of reach of most farmers—apart from the few that manage to access the subsidy programs.

Fortunately, a grey market of unbranded products began to offer drip systems at US$350/ha. Then, one innovative manufacturer introduced a new product labeled "Pepsi"—basically a disposable drip irrigation system consisting of a lateral with holes. At US$90/ha, Pepsi costs a fraction of all other systems.

Source: van Steenbergen 2002.

5.5 THE NEED FOR MAJOR INSTITUTIONAL CHANGES: THE ROLES OF GOVERNMENT, USERS, AND THE PRIVATE SECTOR

Many of the problems of the irrigation sector described in chapter 2 stem from the setup of institutional structures. High fiscal cost, low water productivity, and poor performance of publicly managed schemes result from the ways in which nations have organized the roles and contributions of public and private stakeholders and set incentives for them to deliver. Governments that assume the role of both developer and manager of large-scale irrigation create bureaucratic organizations that have difficulty assuring cost-effective service. Farmers who are faced with unreliable and inequitable water supply and undetermined water rights, and who have no say in how things are run, are reluctant to pay water charges. With the background of the incipient shift in stakeholder roles from top down toward bottom up, discussed in chapter 3, this section reviews in turn the essential role of governments in AWM and the political economy pressures that need to be managed, the role that irrigation farmers can play, and the potential for commercial private sector involvement.

Principles of public intervention in AWM

The "new public management" paradigm is one in which the public sector is a facilitator, developing and enforcing rules for the private sector to interact in markets (World Bank forthcoming). The role of the public sector in AWM differs from this paradigm in several ways. First, water—and environmental protection—are public goods that require public intervention in allocation and management. Second, the huge investments in hydraulic infrastructure are beyond the capacity of the private sector in most countries. Finally, AWM is a critical activity for the overriding public policy objectives of food security and poverty reduction. Therefore, the role of the public sector in AWM is likely to be broader in most countries than the strict paradigm would allow. Although every nation has different goals, policies, and history, specific tasks in relation to agricultural water are assigned to the public sector. These tasks are discussed below.

Policy and institutional options. Governments should be responsible for *core public sector functions* related to AWM. Governments should generally manage and regulate the water resource allocation and governance framework, protecting land and water rights, setting the incentives and institutional structure, and ensuring integrated management of the resource for the optimal welfare of society (including basin management and holistic approaches). Governments should also be responsible for public policy formulation, for strategies for public interventions and for programs of public investment. Governments should also finance core public goods in the AWM sector; including integrated and sustainable water resources management, environmental protection and the management of externalities, research and technology transfer, and rural infrastructure such as farm-to-market roads.

In addition to these core functions, governments should also carry out some transitional tasks in AWM, basically to correct market failure. Examples of where governments should intervene in AWM include the following:

- *Poverty reduction.* Governments should typically (a) adopt pro-poor investment programs, such as cost-sharing investments in small and medium irrigation; (b) affect market prices to allow the poor greater access; and (c) pursue inclusive governance programs to improve the participation of the poor and other excluded groups such as women in WUAs.
- *Water prices setting.* The role of governments in setting prices for large-scale irrigation water service and other key determinants of the cost of water such as energy prices was discussed previously. Governments also have a broader role in trying to ensure that water prices reflect long-term societal values embedded in the opportunity cost.
- *Financial market failures.* Much investment in irrigation is long term and slow yielding, and financial markets fail to accept the levels of risk and financial exposure required over long periods. Governments typically need to correct these failures through guarantees or direct financial intervention.

- *Product market failures.* Irrigated farmers can face very high costs when markets fail. For example, fertilizers are a key accompaniment to AWM in achieving higher productivity, yet in some Sub-Saharan Africa countries their cost is prohibitive, typically greater than two times more than in OECD countries, Asia, and Latin America. Inefficient markets and high transport costs are to blame. There is clearly a role for the state in input market development, as there is in output markets, too (World Bank forthcoming).

Within large-scale irrigation, governments should clarify their responsibilities in financing and managing the different hydraulic components of the systems. In large-scale irrigation, at a minimum, governments have to ensure the public good aspects of water resources management, equity among users, targeting of poverty reduction, water pricing, and mobilization of the required financial resources. Thus, governments should in general take responsibility for the funding and execution of construction, rehabilitation, improvement, and operation and maintenance of the headworks and main infrastructure. Downstream works—secondary canals and below—that directly serve users and that are of a scale that may be financed and operated by users should have the maximum financial contribution from users and maximum user involvement in management. In the future, models for partnership or private development may emerge (see below), but in the interim, governments must determine and assign roles and responsibilities between the public sector and users.

Once these responsibilities are well defined and formalized, governments need to establish institutional structures that have incentives to deliver water service that responds to demand, that are fiscally efficient, and that maximize water productivity. Although there can be no blueprint, the following lessons of experience are important:

- Apply the principle of subsidiarity, that is, decentralize decisions and responsibility to the lowest possible level, creating scheme financial autonomy and accountability.
- Maximize organized user involvement in decisions and financing, bringing demand-driven incentives to efficiency and a spirit of accountable ownership to farmer participation and cost sharing.
- Target cost-effective service delivery and water productivity with benchmarking to track performance and build modernization programs that target the most cost effective improvements (see chapter 6).

Future public investment in large-scale irrigation should be guided by the lessons of experience. Where high returns and poverty-reducing irrigation projects are feasible, as in Ethiopia, for example, governments should take the initiative based on clear criteria for their involvement, such as:

- The allocation of risks and costs has to be fiscally efficient. For example, depending on circumstances, market- or user-financed options should be employed wherever possible.
- Schemes should only go ahead if the benefit stream is adequate to allow users to pay their assigned shares of capital and recurring costs.
- Management should be decentralized and accountable to all stakeholders.
- Users in organized institutions should be involved as partners at all stages, from identification onward (see below).

In moving toward new institutional structures based on revised allocations of responsibilities between stakeholders, governments have to deal with political economy considerations. Stakeholder roles and institutions are shaped in part by the political economy of each nation. Governments target multiple growth and equity objectives and are influenced by various constituencies in the weights they assign to each objective and in the way in which they manage trade-offs. Market-driven growth in the irrigation sector may conflict with equity and poverty reduction objectives. Existing property rights may inhibit more equitable water allocation. Bureaucracies may be well entrenched, often with Byzantine complexities, and bureaucratic interests and incentives may conflict with efficiency goals. Existing entitlements to subsidies and rents may not provide incentives to efficient service or water productivity. The structure of established interests means that in any change there will be losers as well as winners, and reforms in AWM typically have high political transactions costs. Clearly, there is no one way to manage change, but successful practice includes some common characteristics:

- The use of transparent and inclusive processes to diagnose problems and identify options for change
- The role of champions in leading change and brokering solutions
- Piloting of reforms to make sure benefits outweigh costs
- Building in incentives, including early benefits for "winners" and support measures for "losers."[16]

Participatory management and irrigation transfer can effect significant improvements in the way water is managed.

Institutional reform in large-scale irrigation is being driven from the top and the bottom. The background and dynamic driving the rapid spread of participatory irrigation management and WUAs were discussed in chapters 1 and 3. From the top down, drivers have been government policies for administrative decentralization and reduction of the fiscal burden, and also to some extent, adherence to a new paradigm of participatory inclusive development. From the bottom up, drivers have been farmers' perceptions that participatory approaches could give them more influence over key factors that can improve farmer income and reduce risks—water entitlements, water service delivery, scheme modernization, diversification, and value for money in irrigation service charges (box 5.19). These changes have resulted in a significant shift in thinking about the respective roles of public institutions and other stakeholders. In many countries, reform efforts are focused on transferring varying degrees of responsibility for operating and maintaining irrigation systems to the farmers, organized in WUAs, as ways to decentralize management and involve stakeholders responsibly. As indicated in chapter 3, WUAs operate now in more than 50 countries, involved in operation and maintenance, setting and collecting fees, and so forth.

Key lessons have emerged on participatory irrigation management and irrigation management transfer. More than a decade of experience shows both successes and failures in establishment of WUAs and in their functioning. OED (World Bank 2002a) found that beneficiary involvement has facilitated better system operation and management and cost recovery. For example, irrigation performance in Mexico, Turkey, and Niger improved when governments adopted approaches that empowered stakeholders rather than set cost-recovery goals unilaterally. In general, operation and maintenance have improved when water user groups have financial autonomy and arrange operation and maintenance themselves. One study of 26 irrigation systems in six countries in Asia (ADB/IWMI 2004) found that systems transferred from public to private or semiautonomous management almost invariably perform better in terms of operation and maintenance, productivity, and irrigation charge collection. Participation has, however, not proven to be a panacea, and evaluation of participatory irrigation management has provided useful insights into what not to do (Svendsen, Trava, and Johnson 1997).

Most notably, user involvement has to be correctly sequenced with upstream reform of irrigation agencies. In Nepal's Sunsari Morang Irrigation project, the initial delegation of operation and maintenance to water user groups took place before the reliability of water supplies could be assured. The groups could not contribute to improved water management and they failed to achieve their objectives. Investment in group formation activities was premature and had to be repeated. The limits of participation also have to be recognized. In the Philippines, several groups found the operation and maintenance task too onerous, given their limited access to heavy equipment—and the tasks had to be returned to state management (World Bank 2002a). An area of inherent weakness is that WUAs effectively represent the different perspectives and interests of the involved farmers—men and women, head-end and tail-end irrigators, and commercial and subsistence producers—and a WUA does not resolve existing problems of inequality (Vermillion 2004).

Box 5.19. An Irrigation Farmer's Perspective

Farming is a very risky business. Irrigation farming should be less risky than rainfed farming, but an irrigation farmer still has to deal with multiple risks. Some of these risks are beyond the farmer's control—price risks, for example. Farmers are price takers not price makers, and this creates risks for the farmer's bottom line. To secure a decent income and be secure enough to invest in productivity-enhancing measures, farmers need to be able to manage other risks, including

Water entitlements. A farmer may be an irrigation farmer, but in most cases his or her right to the vital input is not secure. The farmer needs the entitlement to be identified and recorded for protection against the vagaries of irrigation scheme management, the competitive groundwater pumping of neighbors, or the demands of a swelling urban community.

Quality of water service delivery. A farmer has to deal with the risk of unreliable or inequitable water supply. If the risk of poor service is high, the farmer is less likely to invest in good-quality seed and fertilizer. On this, the evidence is clear: the creation of WUAs helps to overcome inequity and unreliability of service.

Inadequate infrastructure. Farmers want to diversify into higher value crops, but often irrigation systems are outdated and poorly constructed, and do not have the flexibility to provide variable flows to meet varying crop water demand. The farmer would like to see irrigation systems modernized around an irrigated agricultural redevelopment plan. The objective should be reliable and flexible water service.

Market risks and farmer profitability. For farmers to be able to diversify, they need access to efficient input and output markets. On the input side, they need skills and knowledge, access to credit, improved seeds, and fertilizer. On the output side, they need ways to reduce their vulnerability in the marketplace such as the commodity groups formed by the WUAs on the Mahaweli System in Sri Lanka. They also need support at the post-harvest and post–farm gate stages, such as better storage and handling, as well as farm-to-market roads and access to market information.

Water charges. Farmers will not pay for inefficient and poor quality water and drainage services—but they will pay for quality service, as long as they can see and influence the cost structure. The debate on cost recovery needs to shift to service, and farmers need to represented in the decision-making councils of the service provider.

Source: Personal communication from Geoff Spencer, World Bank, March 2005.

Policy and institutional options. User participation should be included from the beginning, and at each step of the process. Participation should be continued throughout the whole cycle, including involvement in all aspects of operation and maintenance and of cost recovery; "upstream" involvement in planning and in the investment cycle; and "downstream" involvement in monitoring and evaluation. Women should be involved, not least because their participation has been shown to strengthen the institutional setup (Vermilion 2004).

Rights should be formalized to permit transfer of functions. WUAs need to have legal personality and water use rights. They also need to have rights related to their own organization, such as the right to require water users to pay for the water service and the right to collect and use a service fee. They also need rights related to water management and the operation and maintenance function, such as the right to select service providers and hire or release staff, and the right to determine, supervise, and implement an irrigation service plan. Finally, they need rights related to investment and improvement and the right to make legal contracts and own property (ICID 2000; Vermillion 2004). Vermillion (2004) suggests five essential steps toward participatory irrigation management:

- Create a vision and mobilize support
- Establish an institutional framework for empowered water users' associations
- Rationalize irrigation financing
- Reform the public sector and develop support services
- Implement participatory irrigation management, measure results, and adjust the strategy

Capacity building of WUAs should be planned from the start. In addition to technical training on water management functions assumed, capacity building is essential for internal governance of WUAs and for administrative functions. As the process continues, the support should continue, and may, in addition, cover new areas such as conflict resolution, on-farm water management practices, provision of agricultural inputs, and development of agribusiness and marketing.

Government, service providers, and farmers should share a vision at the outset of what the system should look like when the reforms are completed. Irrigation modernization is a process of change from supply-oriented to service-oriented irrigation. It involves institutional, organizational, and technological changes and transforms a traditional irrigation scheme to a flexible one able to respond to market signals. Experience shows that commitment is required for institutional reform at all levels. The expected roles of the primary parties, including farmers, irrigation service providers, bulk water providers, and government must be clearly understood. There needs to be clear recognition of the fact that irrigation water costs to farmers are generally going to increase, especially in the case of irrigation management transfer and elimination of subsidies. One implication is that institutional reform programs should be initially started in agricultural areas where farmers earn better financial returns and can afford to pay a larger share of the real irrigation costs. In areas where farmers are barely breaking even, it is necessary to work on extension and agricultural production improvement programs in conjunction with irrigation reform programs.

Irrigation management transfer (IMT) should be undertaken only when the conditions are right and should generally be a long-term goal. IMT has been largely successful where farmers have water rights and farms are medium and large scale with good access to output markets (as in Mexico and South Africa). Where there are large and small farmers together, or imbalances between upstream and downstream users, transfer will be difficult because the government wil! not be able to ensure equity. In addition, costs are likely to rise as subsidies are eliminated. There is also some evidence that user associations may be higher-cost operations compared to well-functioning public management. Finally, transferring a scheme out of public management requires clear allocation of financial and operational risk and liability between the parties (Vermillion 2004).

Scaling up to water boards or user federations should be encouraged. Water boards are typically established at the level above tertiary and represent several WUAs. Their advantage is that they can be the counterpart for scheme management at a higher level and contribute to policy and management decisions. In Egypt, water boards may govern the management of irrigation, drainage, and domestic or industrial water supply systems. In irrigation, they are elected by their fellow water users and represent WUAs at the level of the *mesqa* (block or tertiary unit). Government is keen on water boards because it acknowledges that the growing pressure on water resources requires a participatory management approach (Vermillion 2004).

Partnerships between stakeholders should continually evolve in a process integral to system modernization. Involving users in the management of irrigation systems through the creation of associations or the transfer of management to these associations is not an ultimate goal in institutional development but part of a dynamic interaction to constantly improve service and cost effectiveness. Developing countries that completed their transfer programs in the mid-1990s (Mexico and Turkey) have done little since then to modernize schemes and improve performance. By contrast, developed countries show the potential of a continued interaction between partners: in Australia, the United States, France, or Spain, irrigation districts are continually improving service, seeking efficiency gains in operation through better information systems and control structures.

The emerging role for public-private partnerships in irrigation

Faced with the challenges of maintaining high investment rates in irrigation and with improving operation and maintenance and cost recovery, governments have turned to new models of large-scale irrigation management. One model was discussed above—sharing tasks, costs, and risks with WUAs, and ultimately transferring irrigation management. Another model, which may be a complement to participatory irrigation management, is through public-private partnerships (PPPs).

A PPP arrangement is, by definition, a contract between a public client and a private operator. Although these arrangements are in their early days in the irrigation sector, experience in the water supply and sanitation sector has led to the emergence of two basic forms of such contracts: the *public contract*, where the government pays the provider usually a fixed amount, which can be a partial service contract (as in Senegal, where a financially autonomous department of the government's local development operator entered into service contracts for maintenance of infrastructure with Senegalese WUAs), or a comprehensive management contract; and the *public service delegation contract*, such as the contract with CACG/Neste in France—a concession contract for water management over a large area in southwestern France, which includes water demand management, such as monitoring withdrawals, control of the resource-to-demand balance, and crisis negotiations, and where the provider is paid according to the operation results. Delegation contracts can take one of five forms: lease, *affermage*, concession, build-operate-transfer, and divestiture (World Bank 2004e).

Three significant lessons can be gleaned from experience to date in the water supply sector. First, PPP helps the water service to become autonomous and enables it to embark on long-term management improvement. Thus, PPP's first positive impact is the introduction of improved management and a corporate culture. A second lesson is that the arrival of a private operator generally spells the reduction of subsidies and entails increased water charges. These increases could have the beneficial effect of driving conversion to high-value-added crops in place of the staple crops that still dominate. On the negative side, a private operator raising charges could create a bad public image and generate reluctance from government and from civil servants to embark on reform. A third lesson is that efficiency gains can be obtained, but they may be offset by the higher cost of private capital and management inputs.

In irrigation, PPP has made some modest beginnings in providing operation and maintenance services and, in one or two pilot cases, in investment partnerships. A study of 21 cases (World Bank 2004e) found that PPP demand has been mostly a government initiative up to now with service providers more reactive than proactive (for example, Tieshan, China, or Manicoba, Brazil). A majority of PPP contracts have covered operation and maintenance functions, either alone or in accompaniment with private participation in investment. Most contracts were public service delegation contracts. Early results of the effect of private participation in irrigation confirm lessons from the water supply sector: water service improved, prices increased as government subsidies were reduced, and the performance of both government and user associations benefited from interactions with a "professional third party" bringing a businesslike approach.

Policy and institutional options. Governments and WUAs should seek out "professional third-party" partners, preferably under public sector delegation contracts with sharper performance incentives. Although experience is scant, PPP plainly can contribute to improving efficiency in the irrigation sector. PPP can particularly improve professionalism and standards in operations and maintenance, both directly and through emulation and capacity building.

PPP projects should concentrate on addressing risks properly. Although numerous risk-mitigating tools exist, PPP risks remain significantly higher in the irrigation and drainage sector than in the water supply and sanitation sector, due to the specifics of the sector. The strong political and social issues related to water, food, and agricultural production maintain *high country risk. Commercial risks*—especially the *nonrecovery risk*—remain high. *Water-specific risks* are also high in all countries where water is scarce and where agricultural water competes with other uses. The presence of high levels of unmanaged risk drives up costs and makes PPP arrangements less viable. Governments need to work with international agencies, such as the World Bank and the International Finance Corporation, to develop contracts that handle risk efficiently and equitably. Recent experience with PPP financing of investments is reviewed in chapter 6, together with recommendations for further development.

5.6 WOMEN ARE AGRICULTURAL WATER MANAGERS, TOO.

Women are stakeholders in AWM—and a poverty target group. Women are important stakeholders in food production in irrigated and nonirrigated agriculture and in nutrition at the household level. They produce two-thirds of the food in most developing countries, and in Sub-Saharan Africa as much as 80 percent. Women are disproportionately represented in poverty statistics, representing 70 percent of the poor worldwide. Yet, a common view about women and irrigation is that women are not involved in it, it's a men's world. There are few documented examples of irrigation approaches that specifically target gender issues.

The impact of irrigation on women and girls is generally positive (box 5.20). The positive impact is felt primarily through increases in labor opportunities, easier access to water—for example for livestock rearing or vegetable gardens—and reduced burden of water fetching (ICID 2000). Irrigation has also been associated with greater power for women in household decisions, and in greater female school enrollment (Lipton and others 2005).

But there are specific problems of participation and equity for women (box 5.21). In India, Bangladesh, Nepal, Pakistan, and Sri Lanka, there is very low female participation in user associations, despite their high involvement in irrigated agriculture. The gender-specific impacts of AWM are widely disregarded in policy and programs. Even in World Bank projects, which in recent years have given serious attention to gender inclusiveness, OED (World Bank 2002a) found that Bank effectiveness on gender issues in irrigation was the least effective of all parameters examined.

Some initiatives, especially by NGOs, have demonstrated that women's empowerment in irrigation is feasible, particularly when financial, technical, and organizational support is explicitly targeted at women. For example, one NGO in Bangladesh, Proshika Manobik Unnayan Kendra with the support of Grameen Bank, was able to organize groups of women for water sales. Even the cultural constraints against being in the fields at night can

---

Box 5.20. The Beneficial Impact of Irrigation on Women and Girls in Bangladesh

In Bangladesh, the increase in labor opportunities generated by irrigation has been higher for female labor compared with male labor. Two-thirds of women in landless and marginal farmer households reported a higher income through increased wage labor opportunities in irrigated production. Women reported that caring for livestock (primarily their role) became easier with irrigation because it increases water availability for bathing cows in the dry season. Incomes from animal production went up. Additionally, irrigation reduced the general work burden of women because it increases access to water close to home, because water collection is primarily an activity of women. Irrigation also changed labor relationships. Before, women often worked for rich households, receiving food in return. With irrigation, opportunities for income generation such as crop processing, agricultural production, or working as agricultural laborers have increased.

Source: Jordans and Zwarteveen 1997.

Box 5.21. Problems Faced by Women in Irrigation in Nepal

Women on the West Gandak scheme in Nepal face specific problems in watering their fields. One problem is night irrigation, because women are not supposed to be out at night. Women also lack access to the informal and formal forums in which water distribution is discussed and arranged. While 44 percent of poor men and 50 percent of middle-income and better-off male farmers participated in such forums, none of the women did. Women are not informed or invited. "It is not practice for women" to attend either informal meetings, such as those held at public places in the village at night, or formal meetings. Even if women are invited and are interested in attending meetings, their husbands may object to their participation. This leaves women irrigators, more than men, with risky informal arrangements.

Source: Van Koppen et al 2001.

---

be overcome: women's groups have hired pump operators to work at night, so that they do not need to be in the fields themselves.

Policy and institutional options. Women should be systematically consulted, empowered, and closely associated in AWM projects, beginning at the earliest design stage. As active participants in irrigated agriculture in most countries, women can bring their own perspectives and distinct sets of interests in how the water should be managed. For example, they will have an interest in investments in low-cost drip systems that can be used for household gardens; they will be concerned about minimizing the need for nighttime irrigation; and they may be effective in mobilizing the other women of the community to ensure that their husbands pay the irrigation fee on time. Culturally sensitive arrangements need to be made to include women in WUAs. Social justice and equity also point to inclusion: because women's livelihoods are affected by how their irrigation systems are managed, they should be represented in that management. Entry points include the following:

- Planning and program design can use economic and social analysis tools to identify women's specific role in AWM and build in interventions that improve the effectiveness of that role.
- Mechanisms of participation and inclusion can be adapted to increase the effectiveness of women's participation, including
  - promoting flexible, reliable, and regular water distribution arrangements for all, including solutions for night irrigation, if needed;

- ensuring that women water entitlement holders have parity, including both women landowners who already are entitlement holders, and female farm decision makers without land titles;
- actively inviting women for meetings, stimulating them to speak up, and creating awareness among women and their husbands and other farmers about the need to extend membership and attendance to all;
- agreeing on minimum quotas of female membership on WUA boards,
- designing any labor obligations (for example, in-kind contributions for cleaning and maintenance works) to allow women to contribute their fair share.

All these options clearly have to be adapted to local conditions.

5.7 IRRIGATION AND AGRICULTURAL WATER MANAGEMENT INTERVENTIONS SHOULD BE TARGETED MORE AT POVERTY REDUCTION.

Irrigation helps reduce rural poverty. For example, when tube well technology was first introduced, it was hailed as a solution for poor rainfed farmers. Even today in Africa, simple hand or treadle pump technology is changing lives of very poor people. Access to irrigation water reduces the incidence and severity of poverty. Recent evidence (see figure 5.5) shows that incidence of poverty is much lower in irrigated areas than rainfed. Irrigation enables households to improve crop productivity, grow high-value crops, generate high incomes and employment, and earn a higher implicit wage rate. A recent IWMI review of 120 published studies on the "irrigation poverty nexus" shows that cropping intensities are higher for irrigation (111–242 percent) than for rainfed (100–168 percent). Yields are also higher for irrigation (rice yields 3.0–5.5 tons/ha) than for rainfed (rainfed rice yields do not exceed 4.0 tons/ha maximum). Employment and wage rates, too, are higher in irrigated areas, with a 50 percent differential not uncommon (Hussain and Hanjra 2004). This has an impact on income inequality and on poverty rates. The same IWMI study shows that income inequality and poverty rates are consistently lower for irrigated areas; and households with access to irrigation and complementary inputs are less likely to be poor.

How does irrigation reduce poverty? Irrigation reduces poverty through three direct first-round effects: increased food output, higher demand for employment, and higher real incomes. The poor populations affected will include the irrigated producers themselves, poor rural laborers, poor net food purchasers in rural areas, and the urban poor (Lipton and others 2005). The poor with access to irrigated land enjoy higher incomes and employment and there is a positive link between irrigation and labor demand generally. Reliable and adequate irrigation raises employment, and this effect increases with

Figure 5.5. Poverty Headcount in Irrigated and Rainfed Areas (2000–2)

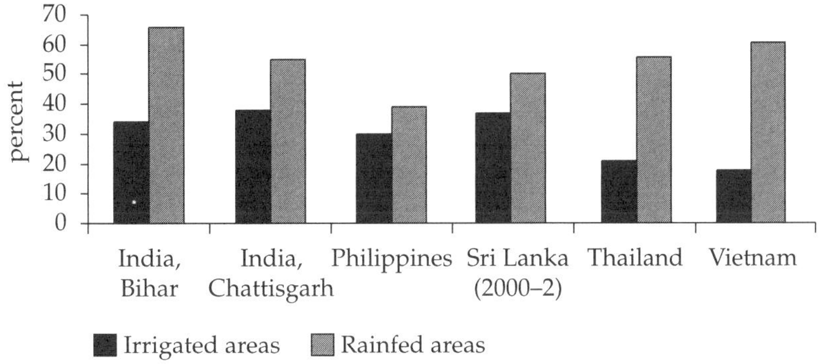

Source: Hussain and Hanjra 2004.

increased cropping intensity. Irrigated areas have more work all year-round. In villages with high intensities of irrigation, employment is almost continuous, creating a continuous flow of cash and food to the household. Other benefits of irrigation to laborers include an increase in daily wage rates, more stable conditions of employment and income, and lower food prices (Lipton and others 2005). Irrigation also has longer-run effects on the poor through a multiplier effect that will drive an increase in nonfarm rural output and employment as the level of rural spending rises. An ADB/IWMI study of 26 schemes across six countries in Asia found that these indirect effects of irrigation were stronger than the direct productivity-related impacts, and that public sector investments in canal irrigation attract private investment in both irrigated agriculture and in the local economy generally (ADB/IWMI 2004). Reduced variability of output, employment, and income will also reduce the vulnerability to risk of the poor. This stability can increase food security and reduce dependence on borrowing (see box 5.22). The better opportunities for crop diversification also reduce risk: a study of the Udawalwe scheme in Sri Lanka showed that household level severity of chronic poverty varies inversely with the crop diversification index (Lipton and and others 2005; Hussain and Hanjra 2004). Social benefits may also accrue. Irrigation has been linked, for example, to such diverse effects as reduced seasonal rural out-migration, and girls' attendance at school (box 5.22).

On large canal irrigation schemes, good management helps reduce poverty. Well-maintained infrastructure improves the impact of irrigation on poverty, while poor maintenance leads to erratic water service and to waterlogging and salinization, which adversely affect the poor tail-enders. ADB/IWMI (2004) found that systems transferred to autonomous man-

---

Box 5.22. Poverty-Reduction Benefits of Irrigation in India

In many parts of India, irrigation by poor families with hand pumps has prevented them from becoming landless. Irrigation can also liberate people from maintaining demeaning social relations such as with money-lenders. For example, poor farmers and landless laborers alike no longer have to "touch the shoes of the rich" in case they have a bad season. Irrigation thus supports self-respecting independence. As demand for labor goes up, the need for laborers and resource-poor farmers to migrate diminishes and may disappear, and families can stay together. It also makes it less difficult to send children to school: in one part of Maharashtra it was possible to send girls to school for the first time.

Source: Lipton and others 2005.

---

agement or with participatory management styles, and with levels of water charges adequate to make schemes sustainable, performed better in delivering water equitably—and so were more pro-poor. Design of water charges can also affect pro-poor outcomes: in Pakistan, farmers are charged according to cropping intensity, which penalizes poor farmers with small plots who tend to double crop. Overall, evidence from the first round of reforms suggests that irrigation management reform benefits the poor, provided that land holdings are fairly equitably distributed (ADB/IWMI 2004).

There may be negative impacts on the poor. Despite the evident benefits of irrigation, many irrigated agricultural systems are still home to large numbers of poor. The ADB/IWMI study found that 38 percent of households on the schemes surveyed were poor, with levels as low as 6 percent in "pro-poor" China and Vietnam, where irrigation has been part of a poverty reduction strategy, and as high as 65 percent in Pakistan, where up to half of households were landless and where land ownership is highly skewed. In addition, irrigation can have direct negative impacts on the poor, and can also have different impacts on different groups. The health and nutrition impacts of irrigation on the poor may be mixed. Access to irrigation may have very positive impacts on nutritional outcomes. However, irrigation may encourage waterborne diseases due to inadequate drainage, particularly the spread of *Anopheles* mosquitoes and schistosomiasis snails, and due to untreated contaminated water. The poor are more vulnerable to such waterborne diseases. When the Karnataka Irrigation Project in India was approved in 1978 the river valley was malaria-free. With the scheme, massive vegetation choked drainage canals, and seepage caused pools of standing water, and malaria returned (World Bank 1994). The poor are also more likely to suffer negative environmental impacts because they frequently face resettlement, and are more likely to be tail-enders and so suffer the consequences of indifferent water services and inadequate drainage. There may also be anti-poor impacts on land and product prices. Higher profitability in irrigated areas may be consolidated into land prices and rents, excluding the poor from access. Higher land costs may increase farm gate prices (Lipton and Litchfield 2003).

Poor people do not necessarily benefit most from irrigation. For example, head- and tail-ender positions affect productivity—and poverty. An IWMI study shows that areas receiving less water per ha have lower productivity and higher poverty rates. Position within a scheme is correlated with poverty incidence. For example, in 10 irrigation areas of Pakistan, wheat yields were found to average 1.7–3.4 tons/ha at the head, but only 1.2–2.9 tons/ha at the tail end (ADB/IWMI 2004; Hussain and Hanjra 2004). Land distribution also affects poverty incidence. Irrigation impact on poverty is highest where landholdings—and therefore water—are equitably distributed. The difference in antipoverty impacts of irrigation improvements between Sri Lanka and Pakistan (box 5.23) was primarily a function of access to land, and to what extent project interventions targeted the poor to correct for initial inequalities (Hussain and Hanjra 2004). Other investments may contribute more to reducing poverty than irrigation. In particular, when the poorest groups are the landless, irrigation may not always be the most efficient poverty-reduction strategy. For example, in China and India, other investments were found to have more favorable benefit-to-cost ratios and higher impact on the poverty headcount than irrigation. These investments included roads, education, research and development, and poverty-targeted rural finance. However, some investments, such as electricity, performed less well than irrigation (World Bank forthcoming).

Policy and institutional options. The findings of the in-depth studies carried out over the last 5 years and summarized above confirm that irrigation does reduce poverty for irrigated farmers, and has the most poverty reducing impact where:

- there is equity in land distribution;
- infrastructure and management are designed with the needs of the poor in mind (for example, equitable governance systems through WUAs);
- water allocation and distribution practices are equitable (head- and tail-ender policies, for instance);
- schemes are well managed and users are involved in management;
- production technology, cropping patterns, and crop diversification are available; and
- support measures such as input supply are in place.

Irrigation also has a broader poverty-reducing impact on the poor through increased employment and lower food prices. Recommendations on how to maximize pro-poor impacts through investments in irrigation and AWM

Box 5.23. Reasons for Difference in Antipoverty Impacts of Irrigation Improvement in Sri Lanka and Pakistan

Improvement of selected systems in Sri Lanka benefited the poor more than other farmers on the schemes, whereas in Pakistan the effect was the opposite—the better off benefited the most. The reasons were that on the Sri Lanka schemes

- inequity in land distribution systems was low (in Pakistan it was very high);
- landlessness was not common (in Pakistan it was high and increasing);
- all irrigation infrastructure was improved uniformly without regard to size of landholding, and the improvements were targeted to the poor (in Pakistan, there was no targeting to the poor); and
- improvement and subsequent governance systems have increased crop water productivity and incomes for the poor (in Pakistan, many of the benefits went to the non-poor).

Source: Hussain and Hanjra 2004.

---

are in chapter 6. Options for poverty reduction through irrigation policy and programs are summarized in this section.

Policy analysis on irrigation and AWM should explicitly examine poverty-reduction aspects, taking into account such factors as the range of technology, because small-scale, low-cost, and labor-intensive irrigation techniques are likely to be important for poverty reduction; distribution of water and land rights, because irrigation investment benefits are largely proportional to access to the factors of production; the objectives and impacts of subsidy to see whether the benefits go principally to the poor; and the social and institutional setup and whether it empowers the poor and women.

Poverty-reduction programs such as Poverty Reduction Strategy Papers should explicitly take into account the role of irrigation and AWM in poverty reduction. Similarly, sector and program monitoring and evaluation (and project economic and financial analysis, see chapter 6), should take poverty aspects into account.

5.8 AGRICULTURAL WATER MANAGEMENT AND THE ENVIRONMENT

As discussed in chapter 4, it is likely that environmental risks and pressures will rise as production expands to meet demand. Developed-country experience shows the range of policy and technology solutions and trade-offs. Many such solutions exist, and more can be sought from technological changes and institutional responses. The challenge is to balance human needs for agricultural products with environmental sustainability and social values, and to find economic ways to reduce negative environmental impacts of agricultural water use. This will require vision, political commitment, institutional change, and the allocation of financial resources. Actions are needed at the global, regional, national, and local levels to develop complementary packages of policy, technical, and economic measures.

Policy and institutional options

Actions are needed at the macroeconomic level. The trade-offs need to be assessed and policies on incentives and on mainstreaming need to be designed. The incentive framework for irrigation and AWM may need to be adjusted to correct market signals to include the value of environmental goods, services, and costs. The most important areas are likely to be reducing energy subsidies to ease pressure on groundwater; eliminating other subsidies to irrigation, and to agriculture generally, that drive environmental degradation; strengthening demand management instruments for agricultural water; and designing equitable but environmentally sensitive incentive structures for multifunctional investments— for example, for recovering the costs of drainage from beneficiaries. As far as possible, the incentive framework should capture economic costs of externalities in prices and should employ indirect measures such as improved irrigation water pricing. The structure could also create positive incentives to sustainability, for example, cost-sharing grants for terrace maintenance and other forms of payment for environmental services. Finally, mainstreaming of environmental concerns into all aspects of water management and agricultural policy is essential (World Bank 2003c).

Environmental concerns should be mainstreamed into research and technological innovation and adoption in AWM. Water scarcity will increasingly drive the technology agenda, and experience has shown that intensification brings substantial environmental risk. Direct measures to reverse water and soil degradation, including watershed management, will be needed, as will indirect measures such as improved irrigation techniques to reduce salinization. Drainage and the reuse of grey water will be important areas for investment (see chapter 6) and the environmental aspects will need to be understood and managed. In rainfed farming situations, environmentally friendly technologies such as no-till methods that reconcile both water management and environmental objectives should be promoted. Governments will need to have proactive environmental policies and investment programs in environmentally sensitive areas where poor people live, with packages of technology, investment, and incentives directed to poor people (FAO 2003d; World Bank 2004b, 2003c).

At the economic level, the main instrument to guide farmers to environmentally friendly practices should be the incentive structure. The use of subsidies should be limited and targeted, for example, to specific pro-poor and market failure situations. Specific support may be directed to poor farmers to enable them to react to environmental and market signals. Direct measures to reverse water and land degradation, including watershed management, may be needed (World Bank 2003c).

Expansion of the irrigated area (estimated at 40 million extra hectares by 2030) can lead to environmental risks. These risks were discussed in chapter 4. In countries where significant expansion is likely and environmental risks exist (such as in some Sub-Saharan African countries), managed expansion would be the best approach, integrating environmental protection for forests and water resource management approaches to reduce pressure on wetlands. Irrigation development plans should specifically address environmental and social issues through the use of strategic environmental assessments, environmental impact assessments and environmental management plans, groundwater management policies, and protection policies for wetlands. Settlement and land development policies should be developed within basin management plans and environmental impacts should be factored in, particularly in Sub-Saharan Africa, where expansion of rainfed areas is likely to be greatest (FAO 2003d; World Bank 2003c).

Risks stemming from intensification of irrigated agriculture were highlighted in chapter 4, including pollution, soil depletion, health effects, waterlogging, and salinization. International experience has shown that the key is to ensure that environmental concerns are "mainstreamed" into the irrigation intensification process. For example, ways to tackle the problem of pesticide pollution in irrigated agriculture are known— the challenge will be to apply them. They include *policies* such as promoting demand for organically grown food; *economic instruments*, for example, taxes and regulation; and *technical improvements*, such as research into smart pesticides, and the use of integrated pest management. Similarly, there are also many tested ways to reduce fertilizer pollution from intensified irrigation. *Policies* on fertilizer can include regulatory measures, such as phasing out low-efficiency fertilizer (ammonium carbonate, for example), as well as promoting public awareness to increase demand for organic products. *Economic measures* can include pollution taxes on mineral fertilizers, or at least removal of subsidies. *Technical measures* on irrigated lands may include the promotion of slow-release fertilizer as part of a balanced irrigation production system (box 5.24); advisory extension services; application of the nutrient budget approach; improved water and nutrient management; promotion of organic farming; and no-till technology (FAO 2003d; Rosegrant, Cai, and Cline 2002b; World Bank 2003c).

Box 5.24. Better Fertilizer Use in China

In China, the world's largest irrigated economy and the world's largest consumer of nitrogen fertilizer, up to half of fertilizer application is lost by volatilization and 5–10 percent by leaching. Better on-farm fertilization management as part of balanced water-soil-nutrient management on irrigated lands, together with regulatory measures and economic incentives for balanced fertilizer use, and technological improvements such as more cost effective slow-release formulations, should help improve the situation.

Controlled-release fertilizers, which become available to plants gradually, improve the efficiency of nutrient uptake by ensuring that nutrients are in the soil in plant-available forms when the crops need them most. Because plants receive the right nutrients at critical stages of development, controlled-release fertilizers can improve root growth, drought tolerance, shoot quality, and flowering while reducing leaching significantly.

Source: FAO 2003d.

The likelihood that agricultural water withdrawals will increase (chapter 4) underlines the threat to environmental flows. Large dams for irrigation and other purposes present risks for upstream and downstream ecosystems. Over-extraction of groundwater for irrigation purposes is another big threat, because aquifers are in some cases not rechargeable or they are exploited at a higher rate than their natural recharge rate. As described in chapter 3, both FAO and IFPRI/IWMI have estimated that increased water will be needed for irrigation in coming years if the world is to be fed. By 2030 one in five developing countries—about 20, most in the Middle East and North Africa and in South Asia—will be suffering actual or impending water scarcity. Stresses on environmental flows are likely to become intense. Environmental allocations will need to be protected—in fact they should increase. General policies that may help with these impacts may include:

- mandating integrated water resources management and basin management approaches, and setting up basin organizations to oversee water allocations;
- developing policies on allocations for environmental uses of water and for reduced pressure on wetlands and other environmental uses of water;
- integrating policy, programs, and incentives for all water-using sectors so that as efficiency of water use improves, institutional mechanisms and incentives ensure that saved water can flow to environmental needs.

For example, "two for one" swaps of recycled wastewater could be brokered in exchange for an agreement by farmers to reduce abstractions.

Some initiatives in developed countries demonstrate that the policy, economic, and technical tools exist to cope with the situation (box 5.25). The main features are a participatory approach involving all stakeholders, a basinwide approach, and proactive and committed central and local government institutions. The challenge is to apply these lessons in developing country situations.

Box 5.25. Environmental Flows and the Living Murray Initiative

The Murray Darling basin is one of Australia's most important natural resources. The basin covers more than 1 million km^2 (14 percent of Australia). The river basin includes 80 percent of Australia's irrigated agriculture and serves a diverse range of functions, including irrigation, drinking water for large urban areas, recreation, fisheries, and ecological functions.

Over the last 10 years, competition for water and declining water quality have created problems. Irrigation has expanded and intensified, drainage water is being reused, and final drainage is now sometimes in sinks away from the river. The rapid rise of salinity had a negative impact on water for domestic use, industry, and agriculture.

Environmentalists and farmers became locked in a fierce debate about declining flows and growing salinity in the river. As a result, the Murray Darling Commission launched the "Living Murray" initiative to take into consideration the values of all stakeholders. This has led to the return of "environmental water" to the river that can flow to the sea and flush the river system to protect fish species and regenerate natural wetlands and vegetation on the floodplains. Now water is being bought from farmers to achieve these objectives.

Source: World Bank 2005b.

6
Investment Options to Promote Agricultural Water Management

Following the discussion of policies and institutions in chapter 5, this chapter examines the role of good investment in driving growth in irrigation and broader agricultural water management (AWM). The chapter looks at the broad range of issues regarding selection, design, and financing of AWM investments.

6.1 THE RANGE OF INVESTMENTS IN IRRIGATION AND AGRICULTURAL WATER MANAGEMENT

In this section, options for decision makers are reviewed for the whole range of investment in AWM, including irrigation and drainage, rainfed agriculture and watershed management, and supply-side investments in water reuse, mobilization of new supplies, and irrigation expansion.

Integrated modernization of existing large-scale irrigation systems

The concept of integrated irrigation modernization embraces all the changes in the irrigation delivery system, in agronomic practices, and in the institutional and incentive structure needed to provide farmers with a sustainable, efficient, and demand-responsive water delivery service. Thus, integrated modernization will require both "hardware" and "software" investments. "Hardware" investments will go beyond the simple rehabilitation of existing systems to include physical improvements to the system such as the right selection of gates and control structures, lining of canals with geosynthetics, construction of interceptor canals and reservoirs, installation of modern information systems, and monitoring and control systems (Supervisory Control and Data Acquisition [SCADA], for instance), as well as on-farm irrigation improvement technologies like drip irrigation. Modernization programs also include a broad range of "software" improvements such as scheme management and institutional structures and on-farm water management practices. Experience shows that modernization focusing only on physical investment (as in the Philippines, see box 6.1) does not address the underlying causes of poor water service and scheme deterioration.

Agenda for decision makers on large-scale irrigation modernization. Integrated rather than single-solution approaches are needed, incorporating physical improvements to the delivery system along with economic, institutional, and agronomic improvements. The best examples of investment in modernization, such as of the Office du Niger in Mali (box 6.1), and the case of Victoria, Australia (box 6.2), include physical upgrading integrated with a range of software improvements designed to ensure that water service responds to farmer needs, is cost effective, and is paid for. Optimization tools have been developed that allow the most cost effective investments to be selected.[16]

Box 6.1. Contrasting Experiences of Modernization

The successful case of Office du Niger in Mali
The modernization of the Office du Niger, which started in the early 1990s, focused on both institutional and technical aspects. The paddy rice processing and marketing functions of the Office du Niger were progressively privatized. The activities of the Office were concentrated around essential functions of irrigation water service, planning, and maintenance. A physical upgrading program focused on improving irrigation water control in the main conveyance and distribution network, and on land leveling. The improved water delivery and land leveling made possible the adoption of rice transplanting techniques and high-yield varieties, resulting in an increase of paddy yields from 1.5 to 6 tons per hectare.

The failed case of the National Irrigation Authority in the Philippines
The National Irrigation Authority (NIA) embarked in the 1990s on a program to transfer operational and management responsibilities to user organizations and to increase irrigation service fees. An investment program intended to promote "modernization" did no more than finance rehabilitation and deferred maintenance and no diagnosis was made to identify the causes of rapid deterioration of the water control infrastructure. As a result, water delivery service did not improve and the schemes continued to deteriorate. Finally, in response to rice shortages and financial problems, the government reduced service fees and reinstated public subsidies to NIA.

Source: IPTRID 2003a.

Box 6.2. Large-Scale Irrigation Modernization in Victoria, Australia

In the state of Victoria in Australia in the 1980s, irrigation suffered from low profitability, aging infrastructure, large public debt, and environmental degradation through salinity and waterlogging. Operation of the complex irrigation channel systems was inflexible, driven from the headworks down. Simple rehabilitation was proposed but analysis of the system revealed opportunities to create a more efficient irrigation system. The roster system requiring the irrigators to take water on a fixed schedule was replaced by a new water-on-order system that allowed farmers to meet their exact crop water requirements. A telemetry system provided real-time management of flows and water levels. The new system allowed leasing of water rights.

Source: IPTRID 2003a.

Along with physical investment, investment in institutional and economic structures is needed to ensure that modernization programs are financially and economically sound. Modernization programs need to generate cash flows to finance not only the modernization program itself, but also subsequent operation and maintenance. Demand management approaches may be needed to encourage efficient water allocation and use (for example, volumetric water charges and quotas, assignment of water rights, development of water markets). Design of the best modernization investments involves users from the beginning, as in the case of Victoria, responding to organized farmer demand across the whole range of modernization investment—in system redesign, in service delivery, and in governance. The demand-driven approach also needs to reflect market conditions, because irrigation modernization requires favorable market incentives (availability of profitable crops and product markets, market information, and infrastructure).

Modernization packages should first be piloted, then implemented in phases. Modernization of delivery service by upgrading existing infrastructure and reforming institutions is a challenging process requiring considerable expertise. These structural and nonstructural improvements should be based on pilot projects to test the applicability of solutions to the physical and social environment of the project and to build the confidence of the users and operators in the new technology and institutions needed to enhance the delivery service. The successful modernization program in Victoria, Australia, for example, was supported by a research program on how to reduce the cost of delivering services and on the technology needed to introduce demand-responsive water service and billing arrangements.

Finally, modernization is unlikely to be a once-for-all event. Programs such as the Jordan Valley scheme are constantly seeking improvement to water service, driven by farmer demand, technological innovation, and market conditions. Where programs are well designed and respond to farmer needs and market opportunities, returns can be high and modernization can improve cost recovery and scheme viability, as in the case of the Office du Niger in Mali.

Improving the performance of small-scale and traditional irrigation systems

A large number of smallholders in Asia, Africa, and the Middle East make their livings from agriculture practiced in small-scale and traditional irrigation systems. Often, these practices are based on community-constructed water diversion and conveyance systems operated by user-managed institutions. These irrigation systems include

- flood-based irrigation systems, such as flood recession irrigation and spate irrigation;
- spring and shallow-well systems common in Asia and in the Middle East and North Africa;
- the elaborate long-distance groundwater conveyance systems found in Western Asia—the *qanats* of Iran, the *karezes* of Afghanistan and Baluchistan;
- small-scale irrigation perimeters lifting water from existing rivers, as in the Senegal or Niger River valleys;
- run-off, run-on micro-perimeter systems exploiting seasonal rains draining from slopes onto bottom lands, as in the *microhydrauliques* systems of paddy cultivation found in Madagascar and Indonesia;
- water harvesting systems, such as the *khukaba* and the *saliaba* systems of Pakistan; Tunisia's water harvesting systems known as the *meskat*, the *jessour*, and the *mgoud*; and the tanks of Sri Lanka; and
- local market gardening systems, usually fed by shallow wells or local runoff and supplying the fresh produce needs of the locality. In Islamic countries, these gardens may use sullage water from the mosque; in Africa and Asia, they may be "kitchen gardens" tended by women.

Although small-scale systems benefit from sound, locally adapted design and from long-established water management user groups, they remain mostly low yielding. Paddy yields in traditional irrigation in Mali are in the range 1–3 tons/ha, compared to 5–6 tons/ha achieved on the large-scale modern irrigation schemes of the Office du Niger. The reasons are the inherent weaknesses of these often remote smallholder economies: lack of economies of scale, high organizational and transactions costs, lack of access to adapted new varieties and techniques, and poor infrastructure and institutional arrangements for input supply and output marketing. The small-scale and traditional irrigation sector has been largely left out of the significant increases in yield and production registered elsewhere in the irrigation sector.

Despite their limitations, these small-scale agricultural water systems have strengths on which development can be constructed. One strength is that well-developed traditional knowledge allows technical and agronomic improvements to be identified and adapted rapidly. A second strength is the existing social capital, with long-time experience in "user association" activity. Recent initiatives have tried to overcome the inherent problems and build on the infrastructural base of these systems. Improvement programs in socioeconomic environments as diverse as Morocco, Burkina Faso, Madagascar, and Niger have been successful in improving yields and incomes.

Options for decision makers on small-scale and traditional irrigation. How to support irrigation management in what is by definition a private and decentralized sector remains a challenging question. Where government policy is to intervene—cost sharing for the improvement of small-scale systems for the poor is a typical intervention—community-driven approaches and related social fund financing mechanisms or working through NGOs may be the most efficient approach. Other successful institutional approaches include working through farmer organizations and working through the private sector, both agribusiness and input suppliers and equipment dealers (World Bank 2005b).

An element of government cost sharing will be necessary, particularly where infrastructure investments are lumpy. Cost-sharing grants from public programs are a common approach. However, some on-farm investments can be made by farmers, especially if supported by investment funds or microfinance interventions (as in the Niger Private Irrigation Projects, box 6.3). As diversification continues and farmer incomes grow, driven by market forces, the need for government direct investment will diminish. (FAO 2004b).

The agenda should include research for the development of affordable irrigation technologies. In addition to improving traditional methods, downsizing and adapting newer technologies may provide the innovations needed. A worthwhile approach used in Niger (box 6.3) is to develop appropriate new technologies and disseminate them through the market. A recent innovation, the market creation approach to development, tar-

---

Box 6.3. Niger Private Irrigation Promotion Projects 1 and 2

The Niger Private Irrigation Promotion 1 Project (PIP1) created local capacity to manufacture and install smallholder irrigation equipment. In four years 1,268 pumps were sold. The irrigators, who farmed on average less than half a hectare, specialized in market gardening. Many of them were able to double or even triple the area cultivated. On average, these farmers' net income increased from US$159 to US$560 in two years, in a country where the median annual per capita rural income is US$60.

PIP2 builds on the success of PIP1, promoting (a) *advisory services and research* to develop and disseminate simple low-cost irrigation and production technologies for small farmers; (b) *capacity building* to strengthen existing smallholder irrigator organizations and to create new ones; and (c) *irrigation investment* through a matching grant facility, and microfinance to help smallholders and irrigator organizations make the needed investment in irrigation. The project is implemented by a local NGO that was established under the project.

Source: PIP1 Implementation Completion Report (June 2002) and PIP2 Project Appraisal Document (February 2002).

---

gets smallholders with innovative products that are inexpensive and cost-effective and that provide significant improvements over traditional methods. Promising places for developing markets for smallholder irrigation include the Gangetic Basin of Bangladesh, Eastern India, and the Terai region of Nepal; the streams and shallow groundwater areas in Sub-Saharan Africa; hill regions in Asia, including Nepal, China, India, Vietnam, and Myanmar; Guishou and Yunan provinces in China; and Deccan Plateau in India.

In addition to these targeted interventions, governments need to invest in broader rural development— rural infrastructure, market development, financial markets—to reduce the risks and ease the structural constraints that plague smallholders: capital scarcity, low enterprise and risk-taking capacity, and poor market links.

On-farm irrigation management

Farmer adoption of efficient, water-saving technologies has been slow and performance is below potential. Water in developing countries is still applied predominantly under conventional border or basin irrigation with very low efficiencies, and staple crops are cultivated using traditional husbandry methods. As discussed in chapter 4, a series of farmer-managed activities can increase the yield per unit of water applied and bring more income for less water through higher cropping intensities, the cultivation of higher-value crops, and the reduction of costs and waste.

Many technological advances are available to improve water delivery to the plant and to improve on-farm water productivity. Micro-irrigation in particular has enormous potential. Drip irrigation uses 30–50 percent less water than surface irrigation, reduces salinization and waterlogging, and achieves up to 95 percent irrigation efficiency. It can be technically demanding—for example, it requires clean water to prevent clogging. Currently, only about 3.2 million hectares are being irrigated by micro-irrigation techniques, just over 1 percent of the total irrigated area worldwide. Sprinkler irrigation covers only 4 percent of world irrigation area. Scope for expansion is enormous: sprinkler and drip are used on 90 percent of the irrigated area in France, but just 3 percent in China (figure 6.1 and table 6.1). Advanced surface irrigation techniques such as precise land leveling, gated pipes, surge irrigation, or cablegation can also greatly improve on-farm water efficiency. Conjunctive use—typically the complementary use of surface water and groundwater—raises the overall productivity of irrigation systems, extends the area effectively commanded, helps prevent waterlogging, and can reduce drainage needs. As water scarcity grows, it is likely to pay increasingly high returns. Technology is widely available and getting cheaper. On-farm technologies such as piped distribution, drip, and bubbler are widely available, and costs can be as low as US$500/ha (or even less with "grey market" systems). Treadle pumps that can irrigate up to 0.5 ha using family labor cost only US$50–100 (World Bank 2005b).

Figure 6.1. Sprinkler and Drip in Selected Countries

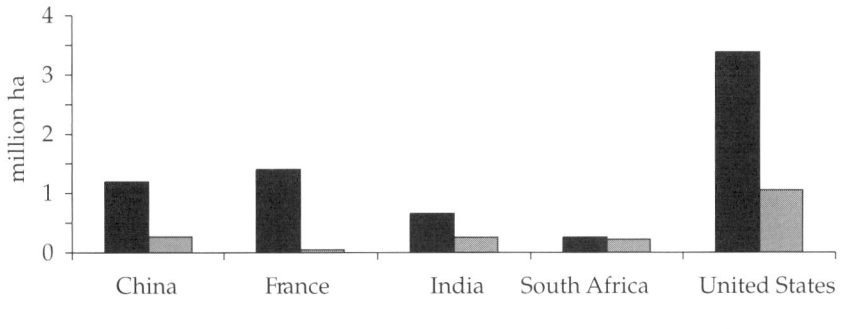

Source: ICID Web site (data refer to different years between 1993 and 2000).

Table 6.1. Coverage of Sprinkler and Drip Irrigation

Country	As percentage of total irrigated area
China	3
France	90
India	2
Jordan	62
South Africa	37
United States	21

Source: ICID Web site (data refer to different years between 1993 and 2000).

Options for decision makers on on-farm irrigation management. Governments should promote farmer involvement in improved technology and practices through the incentive structure, by tightening water charges and energy prices and by encouraging the development of profitable product markets. As discussed earlier, considerable technology is available for on-farm irrigation management, but the challenge is to encourage take-up by farmers. In all approaches to technology development and adoption, motivation of the farmers is vital, and participatory approaches are likely to work best.

Governments have a key role in promoting these investments through the incentive structure. Worldwide experience shows that adoption of efficient technology is greatly hastened by incentives, for example, high water charges volumetrically metered combined with profitable outlets for high-value crops. Countries where both conditions exist have the highest rate of adoption of micro-irrigation. In Jordan, the combination of these factors contributed to two-thirds of farmers adopting drip irrigation (box 6.4). India—one of the world's largest irrigated economies, where water charges are typically low and energy costs are subsidized—has very low adoption rates of micro-irrigation. The message is clear and complements that of chapter 5: governments should promote farmer investment in efficient on-farm water management by a combination of "negative incentives" including higher water charges or energy costs and "positive incentives" such as development of profitable market outlets. Accompanying government rural development investment in road and market infrastructure will also encourage irrigation intensification.

Governments have a role in developing technology and in promoting its adoption through the market and other approaches. The most effective adaptation and dissemination will take place through the market as has happened with drip irrigation in India or treadle pumps in Bangladesh. However, governments facing slow progress can "prime the pump." They should consider investing directly in applied research programs in partnership with the private sector. Farmer investment in improved technology costs money and

---

Box 6.4. Rapid Adoption of Drip Irrigation Technology in Jordan

Water shortage in the Jordan Valley is extreme, and demand management measures have been introduced, including irrigation water quotas and a step tariff that penalizes excess water use. However, farmers also have profitable market outlets for high-value fruit and vegetables. As a result, about two-thirds of the farmers in the Jordan Valley have shifted from surface to drip irrigation over a 10-year period. Farmers have constructed on-farm storage reservoirs to provide the flexibility required for drip irrigation.

Source: Authors.

---

even where the incentive structure is encouraging, farmers may lack the resources. Market-based supplier credit systems have the biggest outreach, and could, if necessary, be supported by public investment through guarantee or refinance mechanisms. Governments can also help the development of market-driven rural financial systems. Governments will need to consider whether subsidies in some form would promote adoption. Chapter 5 reviewed the advantages and disadvantages of subsidies. Where the objective is a vibrant private market, subsidies have to be managed with particular care, but well-designed cost-sharing programs have been successful in promoting water conservation technology. In Tunisia, for example, the National Water Conservation Program (*Programme national d'economie d'eau*) achieved the adoption of water saving technology on close to 200,000 ha (or about 50 percent of the total irrigated area) in five years with matching grants of 30–60 percent depending on farm size and the technology adopted.

Where investment in conjunctive use is appropriate, it should be done as part of an integrated basin plan and in partnership with users. Conjunctive use of surface- and groundwater should be developed as a component of an overall basin-level integrated water resource investment and management plan. It requires coordination between different public institutions and usually a "public-private partnership," because tube wells are generally private. On existing large-scale irrigation schemes, conjunctive use should be part of the overall modernization program because it requires changes in both the hydraulic structures and operating systems to fit new water-use patterns.

Developing drainage

Drainage poses a serious problem. Drainage is land and water management through the processes of managing excess surface water and shallow water tables—by retaining and removing water—with the aim of achieving a balanced mix of economic and social benefits while safeguarding the key ecological functions. As discussed in chapter 2, much of the world's irrigated land suffers from drainage problems, and an estimated 20–30 million ha need improved drainage. The resulting waterlogging and salinity due to rise of water tables and accumulation of salts are reducing water productivity over wide areas and leading to significant social and economic losses for individuals, households, local communities, and countries. However, investment in drainage is usually neglected in developing countries because projects have focused on upstream irrigation and farming (World Bank 2004b, 2005b).

The principal reasons drainage problems are not being addressed today are not only technical but also institutional and economic. The *institutional causes* include first the *lack of an integrated approach*: drainage has been seen as a sector apart whereas it needs to be combined with irrigation in a joint approach to AWM in the context of integrated water resources management. Essentially there should be no investment in irrigation without integrated planning for the drainage counterpart. In integrated planning, the impact of both interventions on the other functions of the resource systems should be carefully examined in consultation with the stakeholders. The current sectoral approach is mirrored in *governance and institutional constraints*: drainage is often a subsidiary task of an irrigation agency. Where they exist, drainage agencies are usually weak, serve only agriculture, and emphasize system development over water management. In fact, drainage needs to be seen within an integrated approach to basin management, as a *multifunctional investment* serving all water sources and users. In line with the integrated approach, best practice shows that *participatory approaches* to drainage work best, although their application to drainage is difficult in practice. Ways to build ownership and management motivation from participatory approaches have not yet been devised, and water user organizations and NGOs rarely deal with drainage. A final constraint on the institutional side is the *legal framework*: the needed enabling legal environment to set up drainage user organizations, to levy fees, and to include the private sector is often either lacking or absent. Recent World Bank experience in the Arab Republic of Egypt and Pakistan identified water boards as the most appropriate type of organization to handle drainage at the secondary canal command levels.

The *economic causes* include the lack of financially sustainable systems: drainage systems constructed are often impaired by low budgets and incomplete cost recovery. Stakeholders do not understand why they should pay for drainage, especially when they were not involved in the planning and decision making. Nonagricultural beneficiaries do not share in costs. *Social constraints* reflect the difficulty of factoring in the poverty reduction aspects. Lack of drainage harms health and social and environmental functions as well as agricultural productivity, mostly affecting the poor—tail-enders who use drainage water for irrigation, the poor who wash in the drains, the downstream population who receive all the pollutants. Yet, the poor have the faintest voice at the decision-making table on drainage. Finally, there are *policy constraints*: drainage is rarely integrated into broader policy platforms such as policies for agriculture, irrigation, and water resources management. For example, in many countries government subsidies promote poor water management and fertilizer pollution, exacerbating the drainage problem, yet there is no counterbalancing policy for dealing with the downstream effects (World Bank 2004b).

Increased investment in drainage is well justified economically. Drainage can improve agricultural productivity and reduce the need for new land development. It can also help with health, flood control, and environmental protection. Drainage projects have generally produced good rates of return and improved farmer incomes (box 6.5), especially where benefits from drainage's "multifunctionality" are factored in. The costs are generally low, ranging from on-farm surface drainage systems at US$100–200 per ha up to US$1,000/ha for pipe drainage in arid areas. A study in India showed that the average cost of creating new irrigated land there was US$6,400 per ha, while the cost of drainage was only US$700–1,000 per ha (World Bank 2004b, 2005b).

Options for decision makers on drainage. Countries need to allocate more resources to drainage investments. As agricultural water use and farming systems intensify, pressures on land and water grow. Both on-site and downstream drainage problems are becoming more acute. Drainage is a partic-

---

**Box 6.5. The Positive Economics of Drainage:
Evidence from Egypt and Pakistan**

An in-depth impact study was undertaken in the National Drainage Project in Egypt. In the project, an intensive network of subsurface pipe drains and open surface drains was constructed in the "old lands" to reduce salinity and overcome waterlogging in the arid environment. A multiyear evaluation of 15 large sample areas established that gross agricultural production increased by US$500–550/ha. The traditional farm annual net income increased by US$375/ha in nonsaline areas and by about US$200/ha in saline areas.

Over the 10-year life of the Mardan Salinity Control and Reclamation Project in Pakistan, crop yields increased between 27 percent and 150 percent.

Source: World Bank 2004b.

ular priority (a) where water is scarce; (b) where there is a productivity problem related to waterlogging or salinization; or (c) where there are other functions of the resource system that can be enhanced by drainage, such as flood protection, pollution control, groundwater and wetland conservation, control of soil erosion and degradation, excessive siltation, saltwater intrusion, or preservation of natural scenery. Thus, each drainage situation is unique and the drainage challenge varies according to the resource system and the prevailing socioeconomic situation: in the water-scarce countries of the Middle East and North Africa the emphasis is on salinity control, in water-abundant monsoon countries of East Asia the emphasis is on flood control. Often drainage fulfils several functions (see box 6.6) (World Bank 2004b).

Drainage needs to be considered within an integrated water resources management framework, not just in terms of agricultural productivity. Drainage is a complex phenomenon with multiple impacts, positive and negative, on the other functions of the resource system. Drainage investments require an integrated approach, addressing all on-site and off-site impacts within the context of a hydrological unit such as a landscape or a basin. Therefore, drainage considerations need to be incorporated systematically into all integrated water resource management and basin plan-

---

Box 6.6. Social and Economic Benefits of Reclaiming Salt-Affected Soils

In Uttar Pradesh, India, about 1.25 million hectares (ha) are barren sodic land and another 1.25 million ha are affected to a lesser extent. The Uttar Pradesh Sodic Land Reclamation project was implemented on 68,400 ha. Cropping intensities increased on all classes of reclaimed land from an average of 61 percent to 220 percent. Yields achieved by project completion were 3 tons/ha for paddy and 2.6 tons/ha for wheat compared with only 0.8 ton/ha from single-cropped paddy fields before reclamation. Labor migration declined from 85 to 50 days per person per year across all families. Real labor wage rates increased by 19 percent for men and 17 percent for women. Unemployment was reduced to 16 percent from 43 percent. These changes triggered significant increase in family incomes. Average per capita income increased by an estimated 87 percent. Average income increased by an estimated 109 percent for families previously considered marginal farmers, by 90 percent for small-scale farmers, and by 59 percent for large-scale farmers. The project was instrumental in alleviating salinity-imposed poverty among area population.

Source: Abdel-Dayem 2005.

ning approaches. In addition, drainage water reuse (see below) needs to be built into irrigation design and management. An integrated approach removes the dominant emphasis on drainage as only a means of mitigating the adverse effects of irrigation, and gives due emphasis to the other direct and indirect benefits, for example, production benefits, the rural development and poverty reduction benefits, and the health and environmental impacts. The capture of the broader benefits enhances economic rates of return and helps justify drainage investments (World Bank 2005b).

Drainage—and integrated water resource management generally—require both participatory approaches and a multistakeholder institutional structure. The institutional counterpart of the integrated water resource management approach is twofold. First, it requires a participatory planning approach involving all stakeholders upstream and downstream, with accompanying awareness programs. Second, it requires a multistakeholder governance and management structure that can arbitrate the trade-offs and seek best compromises between different stakeholders and institutions.

The integrated approach requires first-rate planning tools to take account of the social, economic, and technological aspects. A new methodology (DRAINFRAME, see box 6.7) provides a tool to evaluate all the different functions of the water resource system and match them to the values that society places on them. When used to identify optimal investments in the system, this integrated approach shows not only how drainage can contribute to economic benefit within irrigation systems, but also how drainage can contribute downstream to overall land and water management and to the environment.

Box 6.7. DRAINFRAME

Drainage has to be assessed in the framework of integrated resource management. This allows drainage to be analyzed within the context of a hydrological unit, such as a basin, using an integrated approach and addressing all positive and negative impacts of drainage on-site and off-site. DRAINFRAME, a new methodology, has been developed for this purpose.

DRAINFRAME is an analytic and planning tool that allows drainage water and reuse to be assessed with an integrated management framework. A participatory planning methodology examines all aspects of the resource system and all the stakeholders, and untangles the multiple impacts and costs and benefits, prioritizes investments, and begins to locate benefits and mitigate side effects.

Source: World Bank 2005b.

Drainage investments need meticulous preparation. Best investments are often highly case- and site-specific, and careful design and piloting are required. There have been significant innovations in drainage technology (World Bank 2005b). One is the use of controlled drainage for water table management: land drainage systems often allow water to move too quickly through the soil profile, whereas controlled drainage slows down the loss of moisture and nutrients through the drainage system. A second innovation is the use of evaporation ponds. Costs are low, useful life can be up to half a century, and environmental problems are largely manageable. Finally, biodrainage can be used to remove excess water by using the uptake capacity of vegetation, especially trees (World Bank 2005b, 2004b).

Enhancing water management in rainfed agriculture

As discussed in chapter 4, rainfed farming systems are characterized by poor and variable water availability, prevailing poverty, high levels of vulnerability to risk, and low-yielding technological production packages. Improving water availability and productivity in rainfed agriculture is essential both for global food production and for household food security and poverty reduction. The challenge of rainfed farming is to find solutions that improve incomes, reduce vulnerability, and introduce technical solutions that are accessible without increasing risks.

Options for decision makers on water management in rainfed agriculture. A focused applied research agenda is a priority. At present, small advances are being made in water harvesting, on-farm water and land management, short-cycle varieties, and so forth. But to achieve significant yield increases in rainfed cereals in developing countries will clearly require sustained attention. The International Food Policy Research Institute discusses the need for a breakthrough in water-harvesting systems and advanced farming techniques like precision agriculture, contour plowing, precision land leveling, and no till techniques (Rosegrant, Cai, and Cline 2002b). The work of the research centers of the Consultative Group on International Agricultural Research (CGIAR) on issues of improved water management for rainfed agriculture was discussed in chapter 3. In particular, the International Center for Agricultural Research in Dry Areas (ICARDA) programs on AWM in are focusing on key issues of improving water-harvesting techniques, on developing drought-tolerant and water-efficient germplasm, and on agronomic management of dryland cropping systems. CGIAR work on rainfed farming needs to be reinforced, and adaptation within national research programs improved. In the Republic of Yemen, for example, the dwindling of the groundwater resource has led to a "return to rainfed" and ICARDA has formed a partnership with the national research and technology transfer agency to enhance water management in rainfed areas.

Governments need to invest in both physical improvements and technology transfer. Physical improvements discussed in chapter 4, such as water-harvesting structures and wells for supplementary irrigation, can be financed by governments, for example, on a cost-sharing basis. Public subsidies are justified both by public good aspects such as environmental benefits of soil and water conservation and by poverty reduction rationales. Technology transfer needs to be part of the investment package, because water management, cropping pattern, and crop husbandry improvements are essential components. Packages with "technical integration" of soil, water, agronomy, and so on have been proven to contribute significantly to AWM and poverty reduction where climatic conditions are adverse and resources scant. There may be solutions to even the most difficult technical problems—for example, drought, salinity, and reclamation of sodic soils For example, in China, the loess plateau watershed rehabilitation project located in the Yellow River basin demonstrated, on an area of 1.5 million ha, that investment in soil and water conservation on severely degraded lands can be profitable for farmers (World Bank 2005b).

For all these investments, participatory approaches are needed to test innovations and to get widespread adoption. In addition, downstream investment in roads, agro processing, and market development will be needed (World Bank 2005b).

In rainfed systems, risk predominates, and investment packages must help reduce that risk. Investments in supplementary irrigation, in holistic and integrated watershed management, and in drought management can reduce the risk of uncertain rainfall. However, farmer risks include not only climatic risk and the need for access to reliable technology and water sources but also risks from unstable land tenure and from poorly functioning product and credit markets. The policy, research, and investment agenda has to help rainfed farmers to manage these risks in addition to improving their AWM.

Investment in rainfed farming should be provided within a broader rural development and livelihoods approach. Governments should invest in broader rural development approaches in rainfed areas. Broader development approaches ensure the integration of the technical, social, and market factors needed, and link the development of rainfed farming to both integrated water resource and environmental strategies and to broad agricultural development plans. Thus, governments should invest in integrated rural development programs within integrated watershed or basin management programs. Support to rainfed farming will be most effective where adequate infrastructure and markets already exist or have a great potential to develop.

Improving and scaling up watershed management

Watersheds comprise the slopes within a basin where rain falls and drains into ground- and surface-water channels and reservoirs. As the collectors of water, watersheds play the key role in the hydrological resource system and their management is of critical importance to the maintenance and enhancement of both quantity and quality of the water resource. In addition, watersheds are typically home to a poor population and to diverse economic activities: forestry, rangeland, and hill farming predominate. In general, watershed management challenges have grown stronger in recent years with population pressure and changes in production systems. Unsustainable agricultural uses, such as "slash-and-burn" shifting cultivation, have led to on-site loss of forest cover, to erosion, and to soil fertility decline. Overgrazing of rangeland has led to deterioration of the vegetative cover and to erosion. As a result, on-site production has dropped and downstream sedimentation and flooding have been growing problems (World Bank 2005a, b).

Investing in watershed management is justified both by the on-site conservation and poverty reduction benefits, and by the downstream benefits of improved water resource collection through enhanced infiltration, and of reduced sedimentation and flooding. The mix of public and private benefits makes a market-driven approach difficult, yet the isolated and fragmented nature of investment opportunities and the mix of private and common pool resources are not easily amenable to large-scale, government-led investment programs. Early World Bank experience in watershed management illustrates the problems. Investments in the 1980s adopted a technocratic, top-down approach through expensive construction of erosion control with very limited involvement of local communities. Often subsidies were used as an incentive to participate. There was a lack of collaboration across sectors, and very limited attention to land tenure. The result was scant sustainability once the subsidies ended (World Bank 2005a).

Options for decision makers on watershed management. Future investment in watershed management should adopt participatory approaches, building on existing local institutions and adopting an "action research" farmer-oriented approach to designing investment programs. More than probably any other investment in AWM, watershed investment requires decentralized approaches for planning and service delivery. Most watershed management problems are site-specific, requiring collective decisions by users and a high level of management input and adaptation to local ecological, economic, and social circumstances (World Bank 2005a). Successful investments such as the World Bank–financed Lakhdar Watershed Management Project in Morocco or the Tunisia North West Natural Resources

Management Project have used community-driven development methodologies to identify "win-win" programs that combine production benefits with environmental protection, such as the plantation of fruit and fuelwood trees, and the development of low-cost water harvesting structures. Secure land tenure, a cash crop orientation, and profitability of investments are crucial, and overall attention to incentives is essential. Experience shows that investments like the planting of fruit trees or the adoption of micro-irrigation allow both income improvement and water and soil conservation. Early returns are needed to maintain farmer interest. Thus, a typical investment is a participatory project with a poverty focus aimed at changing land use and boosting incomes through higher-value crops and more sustainable practices, combined with water and land conservation investments. Sustainability is helped if there is a broader investment in rural and human development through improvements to roads, education, diversification, and wider livelihood improvement. A new generation of Bank-financed "integrated" watershed management projects in India, Latin America, and the Middle East and North Africa has resulted in a package of water harvesting, groundwater recharge, environmental protection and vegetative cover, and the development of viable agricultural systems to improve rural livelihoods (World Bank 2005a).

A public cost sharing contribution is justified by the public good and poverty reduction objectives. Best practice tries to minimize subsidy, as high levels of subsidy distort incentives and reduce sustainability (chapter 5). Subsidies are usually needed, however, and typically matching grant approaches have worked best. An alternative is direct payment for environmental services as with the Global Environment Facility project in Central America, which provides incentive payments to farmers who adopt silvo-pastoral techniques on degraded pasturelands (World Bank 2005a, b).

Waste and drainage water reuse for agriculture

Nonconventional water can be a new resource for agriculture. Reflows represent a large potential source of secondhand water. Globally, only about 60 percent of water withdrawn is actually consumed (2,900 Bcm out of 5,190 Bcm)—the rest is returned to the hydrological system and is potentially recoverable in agriculture or other second-round uses. If all this water were recovered, it would be more than three-quarters of the present consumptive use in agriculture (figure 6.2).

Particularly in water-short countries, investment in reuse of treated waste and drainage water can offset water scarcity. Both wastewater reuse and recycled drainage water represent an important AWM investment opportunity. Investment in reuse of low-quality water in agriculture can offset water scarcity and preserve better-quality water for higher-value uses. The

Figure 6.2. Estimated Reflows, 2000

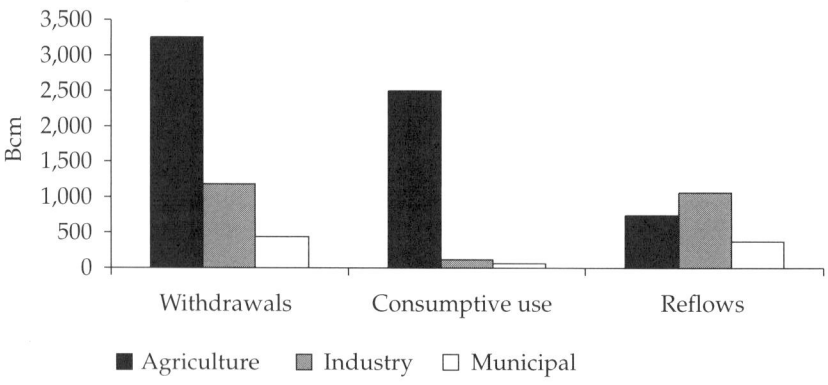

Source: Authors' estimates based on Gleick 1993, among others.

potential for agricultural reuse of treated wastewater is likely to grow as the volume of wastewater from cities is increasing: of the 1 billion people expected to be added to the global population by 2015, 55 percent will live in cities, nearly all in developing countries and, as per capita demand rises, urban water supply will rise even faster than population growth.

At present, systems for recovering and reusing urban wastewater are generally not yet in place in the developing world (figure 6.3.), but as water scarcity grows, investment in treatment and recycling will become more viable. Wastewater offers interesting economic opportunities, particularly in water-scarce basins, because fresh water is likely to be increasingly diverted from agricultural to urban uses. Water reuse in irrigation offers the opportunity to restore some of the water to agriculture and it can even be part of formal agreements, compensating farmers for water they may have given up. It has the advantage of being rich in nutrients. The economic viability is strengthened by the proximity of the resource to urban markets, which usually allows wastewater to be reused for a profitable form of peri-urban agriculture such as market gardening. A further justification for encouraging wastewater reuse is that the downstream economic reuse stimulates environmentally beneficial upstream activity, including management of effluent loads and investment in treatment. Also, reuse in water-scarce areas can preserve better quality water for higher-value uses (World Bank 2005b).

Drainage water is likely to be an important resource, too. Drainage systems collect, evacuate, and dispose of excess surface and subsurface water from cropped fields. Farmers can reuse this drainage water for irrigation, either as a sole source, mixed, or alternated with fresh water from canal or

Figure 6.3. Water Treatment Gaps

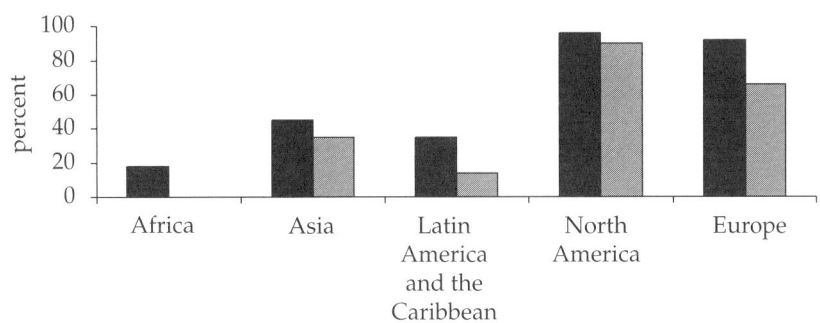

■ In large cities, percentage of population with sanitation
▨ Percentage of sewered wastewater treated to secondary level

Source: World Bank 2005b.

from rainfall. Drainage water is available in large and reliable quantities close to the reuse site and its reuse requires low investment that can be added on to existing schemes. In Egypt, reuse of agricultural drainage water became national policy during the 1980s, and now planned reuse is practiced on 90 percent of the irrigated area, using about 6.5 Bcm or 15 percent of agricultural water use (World Bank 2005b).

Options for decision makers on wastewater reuse. With growing water scarcity in many basins and with the rapid growth of domestic and industrial use, waste water reuse represents an evident area for investment, and one where governments have a necessary role. Governments have to decide wastewater reuse policy and establish the regulatory framework. A series of interrelated policy decisions is necessary prior to investment, regardless of whether the government or its agencies are direct investors. Decisions are needed on the *technical aspects*, because the decision on how to reuse the effluent will determine the decision to invest in intensive (activated sludge, for instance) or extensive (stabilization ponds, perhaps) technologies, or in centralized rather than decentralized systems. The *water resource allocation aspects* have to be designed: decisions on abstractions of water from rural areas and on reinsertion of treated wastewater into agriculture need to be the subject of dialogue and contracts. In particular, if the objective is to reduce or stabilize levels of agricultural water use, some kind of enforceable swap contract has to be agreed on. These arrangements have to be made in an intersectoral fashion because water resources management is typically under the control of one agency, urban water supply and sanitation of another, and agriculture of a third.

On the *environmental side*, a coherent legal and institutional framework is needed, including environmental policies to control contaminants at the source and to reduce waste loads (for example, through application of regulation and the polluter-pays principle). The rules for treatment and reuse must protect human health (box 6.8). Finally, on the *agricultural side*, restricted irrigation and cropping practices will need to be applied, particularly in view of the increased phytosanitary control required in world markets. Governments have a particular role in promoting suitable downstream technology: reuse through surface irrigation, particularly drip but also controlled furrow irrigation, presents less risk of contaminant transmission than does groundwater recharge. Governments also have a role in regulating reuse technology and in promoting appropriate on-farm reuse practices.

Despite the advantages of treated wastewater, important trade-offs need to be managed. In developing countries, the appropriate controls over reuse may be difficult to impose, creating risks to human health and economic activity, particularly if the water is used to grow produce for the export

---

Box 6.8. The Treatment of Wastewater Issues in the Yemen Sana'a Basin Water Management Project

Water availability in Sana'a, the capital city of Yemen, is one of the lowest in the world. Bringing in water from outside the basin is highly uneconomical. The Sana'a Basin Water Management Project aims at increasing the volume and lengthening the useful life of water within the Sana'a basin, by testing many methods of demand and supply management under close monitoring. On the supply side, improved quantity and quality of treated wastewater are crucial. The Sana'a wastewater treatment plant was completed in 2000, but the quality of the treated effluent has been variable. The plant is often bypassed, especially during peak flows, due to energy blackouts, when oil and slaughterhouse waste arrive with sewerage. The plant then releases untreated effluent into the wadi. The effluent, treated or untreated, is used by about 600 farmers to irrigate about 300 ha of crops, including vegetables. This practice poses a high risk of aquifer contamination and endangers the health of both farmers and vegetable consumers. Infiltration from cesspits and inadequate treatment of urban sewerage result in heavy biological contamination in the shallow alluvial aquifer under Sana'a and often cause flooding in parts of the city. Hence, one of the key conditions posed by the Bank for passage from phase one to phase two of the project is compliance with effluent standards by the Sana'a wastewater treatment plant.

Source: World Bank 2005b.

trade. Cost recovery may also be a problem, because neither the previous nor the subsequent user should be responsible for the whole cost of treating the wastewater. The choice between high-quality water and treated wastewater loaded with nutrients also represents a possible trade-off that could be considered, although health problems could pose some restrictions. Finally, meeting the objective of effecting net savings in water-scarce areas by substituting used water for fresh water may be difficult because a contract and regulatory system is hard to implement (World Bank 2005b).

Options for decision makers on drainage water reuse. At the level of investments and institutions, the reuse of drainage water has to be engineered into the system. Drainage water reuse requires a recovery-based loop system that can bring drainage water back into the system. Gravity systems that require low investment can typically be added on to existing systems, consisting of pumps to lift water from drains to canals, construction of mixing basins, and so forth. Farmers may invest themselves and pump water from the drains but this has to be within a framework of guidelines set by sectoral management. Mechanisms are needed to monitor the volume and quality of drainage water and to provide management information for decision making. Programs for reuse of drainage water need to be developed in association with users, and need to be the subject of explicit water entitlements in the same way as fresh canal water. Farmer awareness and training for managing the relatively saline water is essential. These features characterize successful programs such as that in Egypt, which has perhaps the most advanced national system, reusing over 10 percent of the annual freshwater withdrawal without deterioration of the salt balance (World Bank 2005b).

As discussed above, drainage is a complex phenomenon with multiple impacts, positive and negative, on other functions of the resource system, resulting in a need for an integrated water resource management approach. Drainage water reuse has to be assessed at the level of overall basin efficiency and socioeconomic benefit. A particular issue is the downstream environmental effect: there may be less salt discharged but reduced return flow to watercourses (World Bank 2004b, 2005b; see also box 5.25 Environmental Flows and the Living Murray Initiative).

A legal and regulatory framework is needed to control drainage water reuse. This framework would include (a) regulation of water quality, particularly salt content and agricultural chemical residues that may have an impact on productivity; and (b) protection of human health—reintroducing drainage water into the hydraulic system may also hold some dangers for human health (World Bank 2005b).

For drainage water, there are trade-offs that need to be managed. Quality problems need careful control, or salts and contaminants will build up in the soil profile. Reuse may also reduce environmental flows, hence there is a need to examine reuse in an overall basin context.

Expanding irrigation and mobilizing new water supplies

As discussed in chapter 4, FAO projects that to feed the world, about 40 million new ha of irrigated land need to be brought into production between 1997 and 2030, an 18 percent increase to the existing stock. Although the dimensions and location of this expansion can be debated, there is generally an understanding that some irrigation expansion and new water resource withdrawals will be needed to meet the food and energy needs of the world in coming years, despite the environmental and social risks (FAO 2003a, 2003c). For these reasons, the World Bank's *Water Resources Sector Strategy* (*WRSS*) argues in favor of more investment in infrastructure: "many (developing countries) have stocks of water infrastructure that are much smaller than those of climatically similar industrial countries. There are, accordingly, major needs for priority water infrastructure to be developed following best practice from a technical, economic, social and environmental perspective" (World Bank 2004e, p. 24). Some countries have scant water harnessed or stored compared to their population—Ethiopia, for example (see chapter 4).

Dams must be studied at the overall basin hydrological and welfare level. There will be less scope for the large dams that store water over years and transfer from areas of high rainfall to dry areas. The environmental and social consequences of these dams will continue to be contested, and it is likely that nations will construct relatively few of them. There could be more investment in the construction of small dams to meet interseasonal deficits and harvest runoff locally, provided these dams contribute to basin efficiency. Small dams do not raise social and environmental problems on the same scale as large dams.

Numerous conditions govern success in large-scale irrigation development. There has been some successful investment in large-scale irrigation expansion in recent years, notably in Latin America, by both public and private sectors. In Brazil, public investment has successfully blazed the trail for private investment (see box 6.9). The key factors of success in Brazil were assured water supply, farmer commitment, and profitable markets, but many other technical, social, and economic factors need to be taken into account.

Options for decision makers on irrigation expansion and new water withdrawals. First, governments will need to consider whether public financing of new dams and large-scale irrigation expansion is the best investment they can make in AWM. Given the risks and costs involved in new diversions and expansion of the irrigated area, priority should be given to improving exist-

---

Box 6.9. Succeeding in New Irrigation Development in the Brazilian Semi-Arid Region

A recent study in Brazil of 11 public schemes covering 85,000 ha found that four were extremely successful, with positive net present values and high rates of return (above 16 percent). Others were less successful in economic terms, with negative net present values and economic rates of return in the range of 8–15 percent. All schemes were highly effective in creating jobs, at only one-eighth of the average cost in other sectors (US$5,000–6,000 compared with US$44,000). Direct job creation was estimated at 200,000, and total job creation all along the value chain at 500,000. The poverty index in the areas where the schemes were sited was 9 percent lower than in the northeast of Brazil as a whole, and the rural economy in those areas grew at a rate 2.5 times faster than in comparable areas without schemes (6.4 percent compared with 2.5 percent annually). The schemes were successful in promoting regional development, in stemming out-migration and in contributing to increased urbanization. Each increment of 1 percent of rural GDP in the scheme areas corresponded to a 1 percent increase in urban GDP, demonstrating the interactive multiplier between rural and urban growth.

The study analyzed the reasons for the success of the "extremely successful" schemes: reliable water supply, titled land parcels, entrepreneurial producers, available services and markets, and adequate transport infrastructure. The unsuccessful schemes lacked some of these factors—in particular, they often lacked adequate water resources, or were isolated from market centers. The study found that the engineering component was the easier part of the project, while the most difficult part was ensuring farmers' successful participation in production and marketing. The study concluded by recommending the promotion of irrigated agriculture as a key strategy for regional development and poverty alleviation.

Source: World Bank 2004a.

---

ing irrigation. In many countries, returns to other investments in AWM may be higher than to expansion (box 6.10) and the best investment in many cases may be in scheme completion and modernization, in promoting on-farm water use efficiency, and in improving the incentive and institutional structure.

New irrigation projects need to be conceived within an integrated water resource management framework, taking into account the social, economic, and environmental impacts of new resource development. The institutional structure of basin organizations and the tools for water resources, social, and environmental assessment are being developed in many countries. They

---

Box 6.10. Investments in Water Diversion and Irrigation Completion in Iran

Iran continues to invest heavily in water diversion. Investment in water in 2003 totaled 13.5 trillion Iranian rials (US$1.7 billion), 2 percent of GDP, with two-thirds going to the major dam building program and to irrigation development. In 2004, Iran had 83 dams under construction with a total additional capacity of 30 Bcm, which will more than double the current 25 Bcm capacity. Investment in irrigation development has lagged behind, and there has been relatively little investment in environmental programs such as watershed management and downstream drainage. The result of the slow pace of irrigation development is an increasingly unbalanced development of hydraulic resources. Only 13 percent of the irrigable command area under existing structures is fully developed for irrigation. Completion of existing surface schemes attracts much higher rates of return (up to 20 percent), than beginning new schemes (less than 5 percent). For example, the recent Bank-supported Irrigation Improvement Project, which basically completed half-built schemes, had high rates of return—(17 percent, with two of the four schemes having rates of return of over 20 percent, very high for an irrigation investment). The lessons are that the returns to both completion of existing projects and to agricultural intensification are high. A shift from financing new structures to more emphasis on completing existing schemes would be within the fiscal capability of the government and would bring higher rates of return on investment, reflected in higher incomes and more employment.

Source: World Bank 2004d.

---

will need strengthening and will need the backup of the legal and regulatory framework. A particular concern will be the maintenance of environmental flows, for which workable methodologies have been developed.

Recent improvements in the handling of irrigation expansion proposals need to continue, including transparency and participation in reviewing proposals; review of all alternatives; and rigorous technical, economic, and environmental analysis using instruments for assessing and mitigating social and environmental impacts and optimizing benefits. The unit costs of development of irrigation have risen in recent years, while low agricultural commodity prices have depressed rates of return to irrigation. In view of the rising financial costs and the increasingly apparent environmental costs, infrastructure construction plans will need comprehensive analysis of the costs and benefits, including environmental and social effects. As much as possible, major headwork infrastructure projects should typically be multipurpose, which should improve the economic evaluation of the irrigation component of such projects. In addition, more evolved mechanisms for evaluating the "multifunctional" aspects—the nonproduction costs and benefits of agriculture with regard to biodiversity, scenic and cultural amenity, and recreation—will allow a more balanced view of the economics of irrigation. As discussed in chapter 5, users and other stakeholders and their organizations should participate in the process from its inception.

The financial corollary of low economic returns will need careful analysis. If recurrent costs and a reasonable proportion of the investment costs cannot be paid directly or indirectly from farmers' earnings, schemes—or at least their financial engineering—will need to be rethought. Sustainability is imperative, and "somebody has to pay for it" (chapter 5). If government is targeting a nonagricultural objective such as food security for the urban poor, subsidies may be needed. The flaws of dependency on subsidy are significant (chapter 5) and nations must be well aware of these flaws when thinking through and accepting the long-term financial burden of irrigation schemes that depend on them.

The private sector should be involved through public-private partnerships (PPPs) wherever possible. Recent developments in PPPs (chapter 5) have demonstrated the scope for innovating, although private sector investment in large-scale irrigation for smallholders is likely to remain limited. Based on the record to date, management should be decentralized, financial autonomy should be an interim goal, and wherever possible private involvement in management should be sought. Typically this would be through the water user association route, leading in some cases to irrigation management transfer, although there is some scope for public delegation contracts.

6.2 THE EVOLVING INVESTMENT OPTIONS IN AGRICULTURAL WATER MANAGEMENT

This section reviews the changes in the nature of investment in AWM that are likely and discusses ways of improving investment design and methods of assessing benefits. Options for financing investment in irrigation and for increasing the role of the private sector are discussed.

The nature of investment in agricultural water management will change.

It is clear from the discussion in this report that the future pattern of investment in AWM is likely to evolve from the historic pattern (see table 6.2). In large-scale irrigation, many of the changes discussed in chapter 3—chang-

Table 6.2. Current and Likely Future Investment Patterns in Agricultural Water Management

	Present	Expected future investment emphasis	Accompanying measures needed
Large-scale irrigation	Large public command areas producing low-value staple crops using surface irrigation	Modernization, conjunctive use, market development	Large-scale irrigation governance changes (more decentralization, financial autonomy, user participation)
Small-scale irrigation	Farmer-financed schemes, based on run-of-the-river, small dams, and so on	Water productivity investments	Demand management measures; technology transfer and cost-sharing arrangements for on-farm investment
Groundwater	Irrigation from tube wells, privately financed	Drip, fertigation, protected agriculture, market garden crops	Correct incentive structure; develop groundwater governance systems for sustainability
Rainfed improvements	Low-yield farming, vulnerable to risk	Water harvesting, supplementary irrigation, no-till agriculture	Major research and technology transfer agenda; incentive structure adjustments; targeted cost-sharing programs

Source: Adapted from Winpenny 2005.

ing approaches to water resource management and the environment, changing government roles and governance structures, and a new understanding of how rural growth and poverty reduction happen—will have a profound influence on how investments are selected and designed. As many stakeholders told the Camdessus Panel, most future investment is likely to be in the modernization and efficient management of existing assets. Investment in new hydraulic infrastructure will likely be more selective than in the past. More recourse to external financing is likely. Governments and donors are likely to insist that internal cash generation be a significant criterion in investment choice, so markets, cropping patterns, and cost-sharing arrangements will need careful scrutiny (Cleaver and Gonzalez 2003; Winpenny 2005).

For small-scale irrigation, in particular, the pattern of intensification is likely to continue, with technology transfer and matching grant contributions from government. In groundwater irrigation, the focus will be on increasing water-use efficiency. Governments will contribute demand management measures, principally through the incentive structure (see chapter 5), but will also have a role in sharing costs on supply enhancement (groundwater artificial recharge programs, for instance) and on water-use efficiency investments. In rainfed water management, major investments will be needed in research and technology transfer and in integrated programs that contain significant water management components.

Ensuring good investment analysis is key.

Throughout this report, examples have been cited of undervaluation of the benefits of investing in improved AWM. These undervaluations range from neglect of the benefits of increased water security in conditions of hydrological variability to neglect of the parallel and downstream direct growth induced by AWM to neglect of the broader aspects of irrigation multifunctionality. Where some of these benefits are captured, returns to irrigation investment may be much higher than previously considered. One study (ADB/IWMI 2004) found that the on-site productivity of irrigation water in Pakistan was US$0.04/m^3 (for an explanation of the water productivity measure, see chapter 4) but that this increased to US$0.24/m^3 when other local level benefits were factored in, and to US$0.48/m^3—12 times the on-site benefits—when all quantifiable major national-level economic and social benefits were accounted for. Future economic modeling and project planning will need to use methodologies to fairly capture all on-site and off-site benefits and costs of this nature.

Concern about environmental and social issues in relation to AWM has been intense in recent years. Best-practice approaches to ensure that these issues are properly handled in investment design and implementation have been codified in various international and national regulations. Environmental impact assessments and the resulting environmental management plans are good examples of tools that have been developed to ensure that environmental and social concerns are built into investment design from the start. Chapter 3 discussed World Bank safeguards, which are a leading but far from unique example of sets of rules and tools. They cover such topics related to AWM as environmental assessment, natural habitats, dam safety, cultural property, involuntary resettlement, and international waterways.

The objective of these rules is to integrate social and environmental concerns into decisions about investment design. Properly applied, the rules can help to reduce and mitigate adverse environmental and social impacts. Transparency requirements ensure that stakeholders are consulted. Typical impacts that might be captured in agricultural water investments include downstream and third-party effects from surface and groundwater withdrawals, polluted runoff, and drainage water, or loss of farmland and displacement. Safeguards may be seen as imposing a cost on investments. However, more and more countries are adopting the procedures, recognizing that properly applied rules should, in fact, improve investment quality and ownership (World Bank 2005b).

Designing investments in large-scale irrigation

Everywhere in the world, including in developed countries, large-scale irrigation investment is dominated by governments, for two complementary sets of reasons. The first set stems from the fact that large-scale irrigation, even though its principal objective is private benefit of farmers, has many public good characteristics; the second concerns the fact that private financing is not forthcoming. Water resources allocation and sustainable management are public services that can only be performed by government—classic public goods. Large-scale irrigation is also often part of a larger multisectoral project that may provide multiple services, including water supply, flood management, and hydropower in addition to irrigation. Similarly, large-scale irrigation generates multiple externalities that cannot easily be internalized, such as downstream effects, waterlogging, and salinity outside the scheme. Exclusion is not always possible in cases such as drainage or flood protection systems. Finally, large-scale irrigation systems require some land acquisition and rights of way.

Public good aspects could, in principle, be handled through the regulatory framework, although this is typically weak in developing countries. However, governments have also been driven to finance large-scale irrigation because private financing is not forthcoming. Large-scale irrigation has typically been financed by governments because private capital markets cannot provide the financing for long-term, slow-yielding projects. In addition, private investors cannot manage the risks involved in managing long-term exposure in a sector highly susceptible to sovereign policy decisions. These risks include

- *Water scarcity and water demand risks.* The rising levels of water insecurity and more frequent incidence of "water shocks," some associated with climate change (droughts, flood) intensify the perception of risk. In addition, water is a strategic commodity—controlled by sovereign governments—that may be diverted to other uses.
- *Policy risks.* Large-scale irrigation development is often required to meet nonmarket policy objectives such as food security, poverty reduction,

or job creation, and product markets may be distorted and unlikely to yield commercial returns.

- *Financial risks.* Exposures are long, and many projects are unable to pay sufficient financial returns.
- *Commercial risk.* Setting water charges and collecting them have proved difficult almost everywhere.
- *Environmental risks.* All large-scale water projects will be subject to environmental impact assessments, but the possibility of objections or unforeseen impacts down the road brings the risk of financial costs and reputational damage to private investors (SIWI 2004).

If the public sector wants to work with private financing, it must recognize the special nature of these risks and develop packages to mitigate them. The high level of risks translates into investor reluctance and potentially high costs. In India, private investors recently estimated that financial returns on capital invested would have to exceed 20 percent for investment in hydraulic infrastructure to be attractive (SIWI 2004). Returns to investment of this order would make private capital much more expensive than public financing, and drive up water charges. Experience in the hydropower sector where 100 percent private ownership build-operate-own models have been broadly attempted shows that most private projects in hydraulic infrastructure need a high level of public support, and many risks end up migrating back to the public sector. In addition, governments often fail to understand the structure, motivation, and constraints under which commercial partners are operating. This experience in hydropower—and the experience in water supply and sanitation discussed in chapter 5—underline the need to carry out proper identification of risks on either side and to formulate contracts with an equitable sharing of those risks (SIWI 2004; World Bank 2004e).

Public-private partnership in large-scale irrigation investment

Large-scale irrigation projects with high-value crops and commercial farming may provide opportunities for public-private partnerships, such as the recently agreed build-transfer- operate investment contract in Guerdane, Morocco (see box 6.11). Lessons from this promising pilot project were first, that the presence of the International Finance Corporation (IFC) was a key incentive to investors, even though there was no IFC financing; and second, commercial tariffs would have exceeded affordability and the willingness to pay of farmers, so public sector cost sharing was essential to the deal. However—and this is the third lesson—the

---

Box 6.11. Experiences of Public-Private Financing in Large-Scale Irrigation in Morocco

Morocco wished to test PPP arrangements for two large irrigation schemes and invited IFC to examine options. One scheme, Guerdane, was a 10,000 ha irrigation area serving 600 citrus farmers where the groundwater source was running out. The government was prepared to allocate water from the dam complex of Chakoukane-Aoulouz and to cofinance the development of the 60-mile conveyance pipe and distribution structure. In July 2004, the bid was won by a consortium led by Omnium Nord-Africain (ONA), a Moroccan industrial conglomerate, with participation of French and Austrian companies. The consortium will enter into a 30-year concession for the construction, cofinancing, and operation and management of the irrigation network. The project will cost an estimated US$85 million of which the Moroccan government will provide US$50 million, half as loan and half as grant. The water tariff agreed by the consortium is toward the lower limit of the existing cost range of groundwater supply, so farmers will benefit from a cost saving.

    The other scheme, Gharb, presented a very different challenge—a large undeveloped public area (55,000 ha) with some traditional localized irrigation and rainfed farming, in the command area of a new dam. Following new policies, the government was prepared to share development costs 50:50 with the beneficiaries or other investors, but not to shoulder the whole development cost. Could IFC come up with a formula for private participation for both development and operations? The proposal is still under study, but most likely some form of management contract will result.

Source: Personal communication from Imane Akalay, International Financial Corporation, World Bank, March 2005; Authors.

---

public sector contribution had to be pitched not too high to maintain incentives for water saving. Fourth, the competitive bidding process resulted in major savings. Finally, careful design was essential, including the need to set realistic prequalification criteria for bidders (personal communication from Imane Akalay, International Financial Corporation, World Bank, March 2005).

    Among developing countries, private large-scale irrigation development has taken place over a wide area only in Latin America. Even there, governments have played a role in "blazing the trail" for private investment. In Brazil, for example, the government has undertaken large-scale irrigation demonstration projects that have then given the private sector the confidence to invest (box 6.12).

Box 6.12. Public Investment Leads the Way for the Private Sector: Irrigation Development in the Brazilian Semi-Arid Region

In the 1970s, the Brazilian government began to establish public irrigation schemes to settle new farmers in the semi-arid region of the country. These schemes were designed to create jobs and boost exports. They were also expected to serve as development poles that would create a regional dynamic, boosting economic growth and reducing income inequalities. Over three decades, the Brazilian government invested more than US$2 billion to develop 200,000 ha, a cumulative public investment of about US$10,000 per ha. A remarkable effect has been the stimulation of private investment: the development of 360,000 additional ha of private land for irrigation was motivated by the new cropping alternatives, technologies and productive processes validated in the pioneering public schemes. Irrigated agricultural production in the area is now worth US$2 billion annually, including US$170 million of annual fresh fruit exports.

A study of the process drew a powerful lesson—it can take a long time for large-scale irrigation to show positive results, in this case about 10–15 years. There is thus a role for the public sector to undertake demonstration schemes, because the private sector is unlikely to invest with such a long payback period unless the model is shown to be profitable.

Source: World Bank 2004a.

Developing PPP has further potential in the AWM sector. Following the interesting initiative at Guerdane, there is scope to explore other PPP cofinancing arrangements for large-scale irrigation. Already the government of Egypt is studying a PPP model for investment in the East Delta. However, service contracting probably offers the best near-term area for PPP development with the advantage that a third professional partner can improve the efficiency of both state and user organizations. There is scope for contracts ranging from private provision of services (for example, operation and maintenance of pumping stations) to contracting out of scheme management (as may happen in Morocco's Gharb, see box 6.12). Also, other forms of partnership in research and technology development and transfer have potential, as in the case of micro-irrigation equipment in India. The Andhra Pradesh micro-irrigation project used a joint venture between a private firm, NETAFIM, and the government to supply micro-irrigation equipment to 185,000 farmers working a farmed area of about 250,000 ha (World Bank 2005b).

Participation of private investment in large-scale irrigation should be encouraged where possible. Most large-scale irrigation projects will continue to be unsuited for private financing because of the risks discussed above. However, for schemes that are capable of generating adequate cash flow, there are some areas where private financing could be facilitated:

- *Governance reform to create decentralized and financially autonomous agencies.* The Camdessus Panel stressed the scope for direct financing of the subsovereign agencies responsible for actual delivery of water, in line with the worldwide move toward decentralization and financial autonomy. In AWM, this would include irrigation boards and districts, river basin organizations, multipurpose hydropower agencies, and even federations of water users. The advantage of direct financing is transparency and accountability of the agencies. Direct financing creates tough requirements for institutional reform, but this is an essential agenda. Governments need to reform their institutional and financial relationships with relevant subsovereign bodies, and the organizations need to restructure themselves to become financially accountable, creditworthy, and solvent (Winpenny 2005).
- *Role of international financial institutions.* Often even sovereign guarantee may not comfort investors. International financial institutions may use their status to reassure investors and to leverage private financing through cofinancing and guarantees. At Guerdane, IFC did not provide financing but its involvement created investor confidence (see box 6.11).
- *Two part projects.* In many cases, large-scale irrigation projects are part of larger multifunctional water development projects, as with hydropower and irrigation development. In such cases, private sector financing may not be viable for the whole project, yet the public sector may not be willing to develop the project on its own. In these circumstances, a multipurpose project may be divided into public and private elements.
- *Devising appropriate financial instruments.* The development phase of large-scale projects carries a high degree of risk, particularly of cost overrun and delay. A completed project is a more secure investment and could be suited to bond financing. Several Indian states have issued bonds guaranteed by government for financing large-scale irrigation.
- *Matching currencies.* Large-scale irrigation revenues are in local currency. With foreign financing, the government carries the exchange risk. Financing on domestic capital markets would remove this risk.

Steps to increase investment in smallholder agricultural water management

Private investment is the norm for smallholders in small-scale and groundwater irrigation and other forms of AWM. Private sector financing (table 6.3)

Table 6.3. Financing Irrigation and Agricultural Water Management

Type of agricultural water management	Typical financing source		Pointers for development
	For working capital	For investment	
System development and modernization for large-scale irrigation	Irrigation service charges, public subsidy	Subsidies, loans and guarantees, foreign grants and concessional loans, municipal bond issues Cost sharing and PPP contracts	Improve water service and the profitability of farming Decentralize to financially autonomous scheme management PPP arrangements for service delivery and investment
Smallholder	Informal savings groups, cooperative savings and credit arrangements, NGO schemes, money lenders and traders, project-specific credit, micro-finance through local formal intermediaries, and so on	Same as for working capital Special government programs, social funds, and the like	Hire purchase and leasing arrangements Develop legal and regulatory framework for land and water rights as collateral Regulatory framework for rural finance
Commercial	Local commercial banks, specialized agricultural credit agencies, suppliers' credit	Same as for working capital	Leasing, venture capital

Source: Adapted from Winpenny 2005.

is suitable for small irrigation schemes, especially groundwater schemes; for on-farm works; and for equipment in situations where water security exists and high-value crops and market access generate cash flow. If water supply is not assured, or if a low-value cropping pattern is maintained, private finance will be in short supply. Smallholders may also lack collateral for borrowing from the market, because land and water rights are often not secure, defined, or tradable. There is experience in using water rights as collateral for credit, for example in the United States and in Chile, but this would be hard to translate to developing countries where the legal and regulatory frameworks for land and water rights are usually weak (Cleaver and Gonzalez 2003).

Options to increase investment for smallholder irrigation. A market-driven approach is needed that will increase profitability and reduce risk, and so encourage private investment to develop. This requires a pro-market and pro-investment environment, and a good regulatory framework. Specific actions to help improve rural financial services—including credit systems for medium and small farmers and associations or the development of trader credit refinancing; and capital goods hire, purchase, and leasing can be promoted. Risk mitigation instruments such as insurance and forward contracting arrangements would also help. The introduction of tradable land and water rights may also over time help the emergence of private financing.

Governments can support investment programs to help smallholders invest in irrigation technology and to shift to higher-value crops. Ideally, the market will provide financing and the government will be in a facilitating role; however, if market response is too slow, cost sharing approaches or targeted subsidies may be needed (as discussed in chapter 5). In all cases, rural development investments to improve access to markets are essential (Cleaver and Gonzalez 2003).

6.3 FINANCING INVESTMENT IN AGRICULTURAL WATER MANAGEMENT

As discussed in chapter 2, there are no reliable statistics on global irrigation financing (Second World Water Forum 2000). The best estimate is in the range of US$30–35 billion a year for both investment costs and operation and maintenance costs[18] (Cleaver and Gonzalez 2003). Levels of investment in large public schemes can in principle be tracked, because they rely almost entirely on public financing. National government loans and subsidies for AWM have certainly been following a downward trend. Lending by the World Bank and other donor agencies for irrigation and drainage has also declined, from a peak of about US$3 billion annually in the mid-1980s to about US$2 billion by the late 1990s, reflecting lower levels of investment in large-scale irrigation expansion, together with supply-side constraints such as reputational risk over environmental and social impacts. However, no systematic exercise has been carried out to quantify investment spending worldwide, even for large-scale irrigation (Winpenny 2005).

Although official discourse concentrates on public financing and large-scale irrigation, there is a high level of private investment in AWM. In its presentation to the World Panel on Financing Water Infrastructure (called the Camdessus Panel after its chair, Michel Camdessus), FAO estimated that private investment fully finances 20 percent of the world's irrigated area, and provides about half of the investment for the remainder (Cleaver and Gonzalez 2003). These rough estimates imply that private investment could account for about half of total agricultural water investment financing. The figures coincide with the pattern of management of irrigation worldwide, where it is estimated that almost half the global irrigated area is under private management (table 6.4). Trends in private financing are hard to detect, because much of the investment is incremental and small-scale on-farm investment or private water resource development. Foreign investment in "plantation"-type AWM projects has dwindled, as vertically integrated multinational companies have sought higher value added and lower risks in other parts of the value chain, such as food processing and distribution. However, privatization of estates in many countries has drawn investment by national entrepreneurs, and in many countries in Latin America, agribusiness enterprises are investing in irrigated agriculture on a large scale (Winpenny 2005).

Table 6.4. Types of Management in Irrigation

Region	Total Million ha	Public agency managed Million ha	%	Farmer or privately managed Million ha	%	Jointly managed Million ha	%	No data %
East Asia and Pacific	71.8	34.7	48.3	34.5	48.1	0.5	0.7	2.9
Europe and Central Asia	31.6	23.0	72.8	2.6	8.2	0.1	0.3	18.7
Latin America and the Caribbean	18.4	4.8	26.1	3.3	17.9	0.0	0.0	56.0
Middle East and North Africa	20.3	5.2	25.6	1.3	6.4	0.0	0.0	68.0
South Asia	73.7	31.6	42.9	32.4	44.0	0.0	0.0	13.2
Sub-Saharan Africa	6.2	0.8	12.9	2.2	35.5	0.2	3.2	48.4
Total developing countries	**222.0**	**100.1**	**45.1**	**76.3**	**34.4**	**0.8**	**0.4**	**20.2**

Source: Van Vuren and Mastenbroek 2000.

In its submission to the Camdessus Panel, the World Bank estimated that future financing needs over the next 20 years would be up to about twice present levels, that is, an expected annual financing need of about US$4 billion. This projection was based on preliminary assumptions for the period 2005–2025 that, in addition to irrigation and drainage projects in new areas, about 80 million ha of irrigated land would be modernized and 60 million ha would be rehabilitated, 30 million ha of degraded lands would be reclaimed, and small and traditional irrigation schemes would be improved on 50 million ha.

Options for decision makers on financing investment in AWM. The time is right to increase financing investments in AWM. There has been a reluctance to invest because of poor performance, especially in the large-scale schemes. Social and environmental concerns have played a part, too, in deterring investment. However, the time may be right for a turnaround. The ceaseless pressure of rising demand for irrigated agricultural products requires higher levels of investment. As discussed earlier, meeting demand will require investment in modernization and intensification all across the irrigation and drainage sector, and also, as discussed above, extension of the irrigated area. At the same time, some factors currently deterring investment are changing. Rates of return are likely to pick up in the coming years as investment and management models that raise water productivity become more widely adopted, and as irrigated agriculture links into high-value markets. Instruments for analyzing and managing social and environmental problems are now better developed, and governments and other stakeholders are increasingly committed to financing investments according to transparent rules acceptable to society as a whole. Improved economic evaluation methodologies will capture benefits not previously factored in, and climate change and hydrological variability will underwrite the case for more investment financing of AWM infrastructure. The approach proposed in *WRSS* to support large investments in infrastructure of all scales and to invest simultaneously and heavily in management solutions has gained widespread acceptance (World Bank 2004e). Thus, it is likely that the scene is set for an increase in investment levels.

An exercise is currently underway through a joint project of the World Bank, the Global Water Partnership, and the World Water Council to prepare a comprehensive approach to financing agricultural water management, to be presented to the Fourth World Water Forum in Mexico in 2006. As part of this exercise, a financing framework of various types of AWM is being prepared along the lines of table 6.5.

The role of public financing

Chapter 5 proposed some rules of engagement for the public sector in AWM: direct involvement in core tasks of public policy and governance; inter-

Table 6.5. Typical Areas for Public Financing and Interventions in Agricultural Water Management

Public interventions to stimulate market supply and demand	Public interventions in core public goods	Public investment to correct market failure
Technology adoption measures – technology transfer – cost sharing on innovative technology Large-scale irrigation management – initial modernization – headworks and scheme management (declining share basis) – water user association creation – operation and maintenance (declining share basis) Pro-poor measures – cost-sharing investments in small-scale irrigation, watershed management, and rainfed agriculture	Technology adoption measures – research Infrastructure development – farm-to-market roads Water resources management services – basin management services – water use incentive structure – groundwater regulation – climate change and other water resource monitoring Environmental protection services	Financial market development, PPP contracts, and risk management Product market development Trade facilitation

Source: Adapted from World Bank 2005a.

ventions to correct market failures, particularly to protect the poor, to align the incentive structure with public policy goals and to promote the development of financial and product markets; and decentralization and user and private participation wherever possible. Chapter 5 also underlined three characteristics of AWM that define a particular role for the state that is often larger than the new public management paradigm would allow: the public good nature of water resource management and environmental protection, the huge size of hydraulic investments involved, and the importance of AWM for the overriding public policy objectives of food security and poverty reduction.

Within these rules of engagement, governments will need to define priorities for public investment. The World Bank's *Agricultural Growth for the Poor* (World Bank 2005a) recommends that public resources should be targeted at stimulating private investment, maximizing productivity growth, and favoring the poor. Based on these principles, the book suggests a checklist of criteria for public intervention: (a) one time or time limited expenditures to jump start self sustaining private investment; (b) coverage only of transactions costs to avoid distorting underlying long-run equilibrium prices; (c) partnership wherever possible with the private sector; (d) institutional solutions introduced should be subsequently maintainable by the private sector; and (e) an exit strategy under which the state should withdraw to a regulatory role in due course. The book then suggests a framework of entry points, which is a useful tool for defining areas for public financing of AWM in line with the proposed rules of engagement. Table 6.5 sets out a range of typical options for public financing of AWM.

6.4 STRENGTHENING THE POVERTY-REDUCTION FOCUS OF AGRICULTURAL WATER MANAGEMENT INVESTMENTS

Poverty and AWM are linked. Improved irrigation and AWM are central to poverty reduction. Some 70 percent of the world's poor live in rural areas, and most of them are dependent on agriculture. Chapter 5 discussed the ways in which irrigation reduces poverty, both through direct impacts on irrigators and through broader impacts on the poor due to increased employment and lower food prices. The vast majority of the rural poor do not have access to a controlled water source. They live on marginal lands or on drylands. Their technological options for improved water management are limited, and they face high risks from rainfall variations. The poor are also exceptionally vulnerable to drought, floods, effluent discharge, aquifer depletion, waterlogging, salinization, and water quality deterioration. Thus, the key agricultural water challenges for the poor are how to achieve food security, mitigate risk, and improve livelihoods. Improved management of available water therefore has a critical role to play in poverty reduction and food security for the poor. The challenge is most marked in Sub-Saharan Africa, where one-third of farmers simply do not get enough to eat, yet the natural conditions are often not propitious for formal irrigation: 94 percent of the often scant precipitation simply evaporates, compared to a global average of only 63 percent, and development of irrigation schemes is sometimes prohibitively costly.

Irrigation investment can contribute to poverty reduction. From the discussion of poverty in chapter 5, it is possible to identify a number of entry points that show how irrigation can reduce poverty. *Choice of technology* is important. Small-scale, low-cost, and labor-intensive irrigation works, and techniques relying on family labor in their construction and operations are suited to the resource endowment of small farmers. Watershed management and water harvesting technology are typically pro-poor investments, too. Low-cost technology (such as treadle pumps) can be a component of AWM packages directed at the poor. *Distribution of water and land rights* also matters, because irrigation investment benefits are largely proportional to access to the factors of production. Pro-poor irrigation investments therefore have to focus efforts on the smaller farmers and tail-enders, who need to be able to secure access to water in the appropriate quantities and at the appropriate times. Solutions are available: simulations show that canal water reallocation can result in productivity gains for tail-enders without adverse effect on head-end farmers, but institutional changes are needed to effect this (ADB/IWMI 2004; Hussain and Hanjra 2004; Lipton and others 2005).

Institutions such as water user associations that have transparent decision-making procedures and that include small farmers and tail-enders should help equitable access. Systems of water rights and water markets might also help, for example, in access to groundwater. However, rights and markets can be hard to set up, and may be less pro-poor than other systems as they consolidate existing access patterns and exclude new entrants (see chapter 5). Attention has to be paid to *mitigating any negative impacts of irrigation development*, because this usually affects the poor most (chapter 5). In addition to handling resettlement issues, for which a code of practice is now generally accepted, social, health, and environmental impacts need to be considered. Because most irrigation interventions involve the use of subsidy of some kind, care has to be taken to *ensure that the benefits go principally to the poor*. Mechanisms involving community-driven development approaches and social funds can help reach the poor. Finally, *the social and institutional setup* is important—it needs to empower the poor and women.

Agenda for decision makers on poverty reduction in AWM. AWM investments that are the most pro-poor must be preferred. *WRSS* (World Bank 2004e) proposed an analytic framework that can be used to develop a pro-poor AWM investment program as recommended in chapter 5 (see table 6.5). Such a program would give priority to

- pro-poor rainfed agriculture water (and land) investment packages and management programs;
- developing and promoting low-cost irrigation technologies, preferably through the market;
- use of community-driven development and social-fund approaches to AWM investment;
- small-scale irrigation and water conservation investments, which are more pro-poor and characterized by high flexibility and rapid implementation;
- targeting large-scale irrigation investments toward pro-poor "entry points";

- diversification into higher-value irrigation food and cash crops, including home garden and other initiatives involving women.

Investing in AWM for rainfed farmers must be a priority. The range of investments possible in rainfed agriculture, where the poorest are found, was discussed above. Investments in watershed management, drainage, rainfed water harvesting, supplemental irrigation, micro-irrigation, and so forth can have a strong pro-poor impact and help to reduce risk for rainfed farmers. Systematic monitoring and evaluation of poverty interventions and impacts, as recommended in chapter 5, is also a priority.

Large-scale irrigation investment should be targeted toward pro-poor entry points. Public investment in irrigation should be made more pro-poor. Project preparation can explicitly improve poverty reduction impacts, by assessing up front (a) distribution of land and water rights in relation to poverty target groups; (b) impact of changes in yields, output, crop mix, and so forth on different types of irrigation beneficiaries; (c) employment effects, separated into short-run, construction-related effects and longer-run, agricultural-related effects; and (d) social, economic, and environmental effects on surrounding nonirrigated areas. Box 6.13 provides a checklist for improving these pro-poor impacts. Key design improvements are likely to include ensuring the inclusive and equitable nature of institutions, including water user associations, land and water rights, and the incentive structure; specific pro-poor targeting of subsidies and services directed at the smallest farmers and tail-enders; maximization of job creation; and mechanisms for factoring in women's roles and minimizing adverse

Table 6.6. How Agricultural Water Management Interventions Contribute to Poverty Reduction

	Broad interventions	*Poverty targeted interventions*
Interventions involving water resources development and management in general	**1. Broad region-wide water resource interventions** For example, river basin development and aquifer management	**2. Targeted water resource interventions** For example, watershed management in designated areas with poor farmers
Interventions involving AWM directly	**3. Broad water service delivery reforms** For example, reform of irrigation management companies or development of water user associations	**4. Targeted water services** For example, grants for water harvesting or supplementary groundwater irrigation in dry land farming

Source: Authors, adapted from World Bank 2004e.

Box 6.13. A Checklist for Improving the Pro-Poor Impact of Irrigation Projects

Pro-poor policies
- Does the project change land tenure or water rights; and, if so, does it do so in a pro-poor way?
- Do expected increases in yields, marketable surplus, and incomes accrue fairly to poor farmers?
- Does the project try to minimize displacement and resettlement of poor communities by opting for smaller infrastructure?
- Are domestic water supply and sanitation in rural areas included as specific objectives of the irrigation project?
- Are other possible income-generating uses of irrigation water (for example, aquaculture, livestock) enhanced by the project?
- Are complementary services (credit, education, extension, for instance) included in the project and do they particularly target the poor?

Pro-poor technologies
- Is the entry price affordable? Do investment and operation costs of the technologies allow access to poor people?
- Have all available technologies for smallholders been considered in the selection process?
- Are there arrangements for pro-poor research and technology transfer?
- Is drainage needed, especially in poorer areas subject to waterlogging and salinity?

Pro-poor water management
- Are the voices of poor men and women adequately heard in participatory water resources allocation decisions—in selection of the project area, project design, development, and operation?
- Are there in-place mechanisms to facilitate the creation of groups of poor farmers, which can strengthen their cooperative negotiation power and make their access to water rights and other complementary services (micro-finance, for example) easier?
- Is adequate technical and administrative support provided to water users associations, and especially to poor men and women?
- Do cost-recovery arrangements (water pricing) and incentive policies adequately protect the poor (perhaps through block tariffs to protect base water consumption)?
- Are distributional issues, for example, head-ender and tail-ender conflicts, dealt with in an equitable way?

Direct and indirect impacts on the poor
- Does the project generate extensive additional employment, both during construction and during subsequent operations?
- Are environmental impacts that may affect the sustainability of the livelihoods of the poor adequately assessed and dealt with?
- Is water-quality management adequately considered (by safe disposal of drainage water), especially when water is used for drinking purposes?
- Are health impacts (for example, malaria and bilharzias) considered and mitigated to the extent possible by the project?

Source: Adapted from World Bank 2002c and from Lipton and others 2005.

social and environmental impacts on the poor (ADB/IWMI 2004; Lipton and others 2005).

Research needs to focus on the types of technology most appropriate for different types of poor users and poor beneficiaries. Technology that creates demand for labor rather than replacing it is likely to be the most appropriate. A key area is adaptation of technology to the conditions of the poor and getting the technology to market. Excellent recent examples of low-cost irrigation technology include drip kits and treadle pumps. As recommended above, governments and research agencies need to forge partnerships with NGOs and suppliers to accelerate these developments. Socioeconomic research into poverty impacts is also needed to provide more detail on which types of irrigation are of greatest benefit to different types of poor people in different agro-ecological regions and institutional settings.

Notes

Chapter 2

1. However, there are huge variations in resource endowment between countries. Some countries (such as China) have little room for expansion, whereas others (such as Ethiopia) have considerable unharnessed surface resources. There are also important variations between water-abundant and water-scarce countries, between rich and poor countries, and between countries that have already developed much of the needed infrastructure and those that have not. The Water Resources Sector Strategy of the World Bank argues that investment in new storage and diversion infrastructure should continue where economically, environmentally, and socially justified.

2. However, it is likely that irrigation benefits have been systematically understated (see chapter 6).

3. Consumption shares are higher than withdrawal shares because in agriculture about half of withdrawals are consumed in evaporation and evapotranspiration, whereas other uses, such as municipal and industrial uses, return a much higher proportion of their withdrawals to the hydraulic system as waste water and in other forms.

Chapter 3

4. Members comprised FAO, Global Water Partnership, the International Commission on Irrigation and Drainage, the International Federation of Agricultural Producers , the International Water Management Institute, the United Nations Environment Programme, the World Conservation Union , the World Health Organization, World Water Council, and the World Fund for Wildlife.

5. Environmental flows are covered later in this chapter.

6. However, outcomes have not matched expectations. Attention to gender issues was rated by OED as the least effective of all Bank actions in irrigation and drainage (World Bank 2002a). See chapter 5 for a discussion of gender and AWM and for policy recommendations.

7. At each stage of development, management and operational considerations will be vital, and planning for operation and maintenance and agreement on service levels should be included in the process of investment design.

8. A special review of world experience has been carried out in preparation for this report (Vermillion 2004). Its conclusions and recommendations are discussed in chapter 5.

Chapter 4

9. The discussion of future supply and demand is based largely on the projections made by FAO (2003d) and by IFPRI/IWMI (Rosegrant, Cai, and Cline 2002a, 2002b).

10. These two sets of projections, which come to similar conclusions, are not used here as forecasts of what will happen, but as guides to what could happen given the vectors of change analyzed in chapters 2 and 3. The projections and accompanying analyses are used throughout this paper to give the context for discussion of likely trends in agricultural water management and of the policy, institutional, and investment implications.

11. FAO figures are for overall irrigation efficiency, including groundwater sources. In many large-scale systems, current efficiencies are as low as 30 percent.

12. See chapter 6 for more discussion of these technologies.

13. A research project, carried out by Roorkee University, the Water and Land Management Institute of Uttar Pradesh, and the State's Irrigation Department, in collaboration with the International Water Management Institute (IWMI), evaluated this ongoing experiment in large-scale recharge, which is being carried out by the government of Uttar Pradesh (personal communication from Douglas Olson, World Bank, March 2005).

14. In its "business as usual" scenario, IFPRI has the somewhat lower projected share of 57 percent for irrigated agriculture's contribution to increased cereals production. However, both analyses underline the critical role of irrigation in producing commodities and feeding the world in coming decades.

Chapter 5

15. Sur, Umali-Deininger, and Dinar (2002) point out that where environmental externalities are created, the irrigator benefits from a second subsidy because the costs of cleanup are picked up by others.

16. See *Shaping the Future of Water for Agriculture: A Sourcebook for Investment in Agricultural Water Management* (World Bank 2005b) for a full discussion of managing policy and institutional change.

Chapter 6

17. One such tool—the Rapid Assessment Program or RAP—is described in *Shaping the Future of Water for Agriculture: A Sourcebook for Investment in Agricultural Water Management* (World Bank 2005b).

18. Estimate made by the World Panel on Financing Water Infrastructure in *Financing Water for All* (Winpenny 2003).

Bibliography

Abdel-Dayem, S. 2005. "Understanding the Social and Economic Dimensions of Salinity." Presentation at The International Forum on Salinity. Riverside, California, April 25–28.

Abdel-Dayem S., J. Hoevenaars, P. P. Mollinga, W. Scheumann, R. Slootweg, and F. van Steenbergen. 2004. "Reclaiming Drainage: Toward an Integrated Approach." ARD Report 1. World Bank, Washington, DC.

ADB/IWMI (Asian Development Bank/International Water Management Institute). 2004. *Pro-Poor Intervention Strategies in Irrigated Agriculture in Asia: Issues, Lessons and Guidelines.* Colombo, Sri Lanka: IWMI.

Albinson, B., and C. J. Perry. 2002. *Fundamentals of Smallholder Irrigation: The Structured System Concept.* Research Report 58. Colombo, Sri Lanka: IWMI.

Aw, D., and G. Diemer. 2005. "Making a Large Irrigation Scheme Work: A Case Study from Mali." Directions in Development Report, World Bank, Washington, DC.

Barghouti, S., S. Kane, K. Sorby, and A. Mubarik. 2004. "Agricultural Diversification for the Poor: Guidelines for Practitioners." Agricultural and Rural Development Discussion Paper 1, World Bank, Washington, DC.

Barker, R., and F. Molle. 2004. *Evolution of Irrigation in South and Southeast Asia.* Research Report 5. Colombo, Sri Lanka: IWMI.

Bastidas, E. P. 1999. *Gender Issues and Women's Participation in Irrigated Agriculture: The Case of Two Private Irrigation Canals in Carchi, Ecuador.* Research Report 31. Colombo, Sri Lanka: IWMI.

Bosworth, B., G. Cornish, C. Perry, and F. van Steenbergen. 2002. *Water Charging in Irrigated Agriculture: Lessons from the Literature.* Report no. OD 145. UK: HR Wallingford.

Bruinsma, J. 2003. *World Agriculture: Towards 2015/2030. An FAO Perspective.* Rome: Earthscan.

Bucknall, J., I. Klytchnikova, J. Lampietti, M. Lundell, M. Scatasta, and M. Thurman. 2003. *Irrigation in Central Asia: Social, Economic and Environmental Considerations.* Washington, DC: World Bank.

Burton, I., and M. van Aalst. 2004. "Look Before You Leap. A Risk Management Approach for Incorporating Climate Change Adaptation into World Bank Operations." Environment Department Paper No. 100, World Bank, Washington, DC.

Byerlee, D., X. Xinshen Diao, and C. Jackson. 2005. "Contribution of Agricultural Households to National Poverty Reduction in Africa." World Bank, Washington, DC, and DFID, London.

Chakravorty, E. 1997. "The Economic and Environmental Impacts of Irrigation and Drainage in Developing Countries." In *Agriculture and Environment: Perspectives on Sustainable Rural Development*, ed. E. Lutz, H. Binswanger, P. Hazell, and A. McCalla, 271–82. Washington, DC: World Bank.

Christoplos, I. 2004. *Out of Step: Agricultural Policy and Afghan Livelihoods.* Kabul: AREU.

Cleaver, K., and F. Gonzalez. 2003. *Challenges for Financing Irrigation and Drainage.* Agriculture and Rural Development Department. Washington, DC: World Bank.

Cornish, G. A., and C. J. Peery. 2003. *Water Charging in Irrigated Agriculture. Lessons from the Field.* Report no. OD 150. UK: HR Wallingford.

CropLife International. 2004. "Water Matters for Sustainable Agriculture. A Collection of Case Studies." http://www.croplife.ca/english/pdf/resourcecentre/WaterCaseStudies(April%202004).pdf

Davis, R., and R. Hirji, eds. Various years. Water Resources and Environment: Technical Notes Series. Environment Department. Washington, DC: World Bank.

De Fraiture, C., X. Cai, U. Amarasinghe, M. Rosegrant, and D. Molden. 2004. *Does International Cereal Trade Save Water? The Impact of Virtual Water Trade on Global Water Use.* Colombo, Sri Lanka: IWMI.

Delgado, C. L., J. Hopkins, and V. A. Kelly. 1998. *Agricultural Growth Linkages in Sub-Saharan Africa.* Research Report 107. Washington, DC: IFPRI.

Döll P., and S. Siebert. 2000. "A Digital Global Map of Irrigated Areas." *ICID Journal* 49 (2): 55–66.

———. 2002. "Global Modeling of Irrigation Water Requirements." *Water Resources Research* 38 (4): 1037.

Droogers, P. 2002. *Global Irrigated Area Mapping: Overview and Recommendations.* Comprehensive Assessment of Water Management in Agriculture. Working Paper 36. Colombo, Sri Lanka: IWMI.

Easter, K. W., and Y. Liu. 2004. "Cost Recovery and Water Pricing for Irrigation and Drainage Projects." Background report for *Directions in Development. Reengaging in Agricultural Water Management: Challenges, Opportunities, and Trade-Offs.* Washington, DC: World Bank.

FAO (Food and Agriculture Organization). 2003a. *Agriculture, Food and Water.* A contribution to the World Water Development Report. Rome.

―――. 2003b. *A Perspective on Water Control in Southern Africa. Support to Regional Investment Initiatives.* Land and Water Discussion Paper 1. Subregional Office for Southern and East Africa and FAO Land and Water Development Division. Rome.

―――. 2003c. *Summary of Food and Agricultural Statistics 2003.* Rome.

―――. 2003d. *Unlocking the Water Potential of Agriculture.* Rome

―――. 2004a. *Risk Management in Agricultural Water Use.* Rome.

―――. 2004b. *Water for Sustainable Food Production, Poverty Alleviation and Rural Development.* Rome.

Foster, S., and H. Garduño. 2004. "Towards Sustainable Groundwater Resource Use for Irrigated Agriculture on the North China Plain." World Bank, Washington, DC.

Foster, S., and Kemper, K., eds. 2003. *Sustainable Groundwater Management: Concepts and Tools.* GW-Mate Briefing Note Series. Washington, DC: World Bank.

Gleick, P. H., ed. 1993. *Water in Crisis: A Guide to the World's Fresh Water Resources.* New York: Oxford University Press.

Gleick, P. H., A. Singh, and H. Shi. 2001. "Threats to the World's Freshwater Resources." Pacific Institute for Studies in Development, Environment and Security, Oakland, CA.

Gleick, P. H., G. Wolff, E. L. Chalecki, and R. Reyes. 2002. "The New Economy of Water. The Risks and Benefits of Globalization and Privatization of Fresh Water." Pacific Institute for Studies in Development, Environment and Security, Oakland, CA.

Govardhan, D. S. V. 2003. *Judicious Management of Groundwater through Participatory Hydrological Monitoring: A Manual.* Arnhem, The Netherlands: Arcadis Euroconsult.

Groenfeldt, D. 2003. *Training Module on Gender Dimension of Participatory Irrigation Management.* Washington, DC: INPIM.

―――. 2005. "Multifunctionality of Agricultural Water: Looking Beyond Food Production and Ecosystem Services." FAO/Netherlands International Conference on Water for Food and Ecosystems, The Hague, Feb. 1.

GTZ (Federal Ministry for Economic Cooperation and Development) and AfDB, ADB, DfID, DGDC, GTZ, MinBuZa, OECD, UNDP, UNEP, World Bank. 2004. "Poverty and Climate Change: Reducing the Vulnerability of the Poor through Adaptation." GTZ, Berlin.

Heim, F., and C. Abernethy, eds. 1992. *Irrigated Agriculture in Southeast Asia beyond 2000.* Proceedings of a Workshop held at Lkankawi, Malaysia, October 5–9. German Foundation for International Development and International Irrigation Management Institute (DSE).

Hoevenaars, J., and R. Slootweg. 2004. "Rapid Assessment Study Towards Integrated Planning of Irrigation and Drainage in Egypt—Natural

Resources Perspective." Preparation study for the IIIMP project in Mahmoudia, Egypt. IPTRID.

Hussain, I., and M. A. Hanjra. 2004. "Irrigation and Poverty Alleviation: Review of the Empirical Evidence." *Irrigation and Drainage* 53 (1): 1–15.

ICID (International Commission on Irrigation and Drainage). 2000. *Proceedings of the 8th ICID International Drainage Workshop, "Role of Drainage and Challenges in 21st Century."* New Delhi, India, January 31 to February 4.

Inocencio, A., H. Rally, and D. J. Merrey. 2003. "Innovative Approaches to Agricultural Water Use for Improving Food Security in Sub-Saharan Africa." Working paper 55. IWMI, Colombo, Sri Lanka.

IPCC (Intergovernmental Panel of Climate Change). 2001. *Climate Change 2001—Impacts, Adaptation, and Vulnerability.* Cambridge, UK: Cambridge University Press.

IPTRID (International Program for Technology and Research in Irrigation and Drainage). 1999. "Realizing the Value of Irrigation System Maintenance." Issues paper 2, FAO, Rome.

———. 2003a. "The Irrigation Challenge: Increasing Irrigation Contribution to Food Security through Higher Water Productivity from Canal Irrigation Systems." Issue paper 4, FAO, Rome.

———. 2003b. *Supporting Capacity Development for Sustainable Agricultural Water Management.* IPTRID Partnership Programme 2003–2005. Rome: FAO.

———. 2004. *HORTICA. Renforcement des capacites de micro-irrigation pour l'intenification de l'horticulture. Zone des Niayes.* Republique du Senegal. Ministere de l'agriculture, de l'elevage et de l'hydraulique. Direction de l'horticulture. Rome: FAO.

IWMI (International Water Management Institute). 2002. *World Irrigation and Water Statistics 2002.* Colombo, Sri Lanka.

Johansson, R. C. 2000. "Pricing Irrigation Water: A Literature Survey." Policy Research Working Paper 2449. Rural Development Department, World Bank, Washington, DC.

Johnson III, S. H., M. Svendsen, and F. Gonzalez. 2004. "Institutional Reform Options in the Irrigation Sector." Agriculture and Rural Development Discussion Paper 5, World Bank, Washington, DC.

Jordans, E., and M. Zwarteveen. 1997. *A Well of One's Own: Gender Analysis of an Irrigation Program in Bangladesh.* Country Paper No. 1. Colombo, Sri Lanka: IWMI.

Kemper, K. ed. 2004. "Groundwater: From Development to Management." *Hydrogeology Journal* 12 (1): 3–5.

Kikuchi, M., A. Inocencio, M. Tonosaki, A. Maruyama, I. de Jong, and H. Sally. 2005. "Costs of Irrigation Projects: A Comparison of Sub-Saharan Africa and other Developing Regions and Finding Options to Reduce

Costs." Draft background paper for Collaborative Program. IWMI, Pretoria.

Klugman, J. 2002. *A Sourcebook for Poverty Reduction Strategies*. Washington, DC: World Bank.

Lipton, M., J. Litchfield, and J.-M. Faures. 2005. "The Effects of Irrigation on Poverty: A Framework for Analysis." *Journal of Water Policy* 5: 413–427.

Mellor, J. W., and S. Gavian. 1999. *Determinants of Employment Growth in Egypt: The Dominant Role of Agriculture and the Rural Small-Scale Sector.* Study sponsored by the Government of Egypt and USAID. Cairo: USAID Office for Economic Growth.

Molden, D., and C. De Fraiture. 2004. "Investing in Water for Food, Ecosystems and Livelihoods." Discussion Draft. Blue Paper. CGIAR, Stockholm.

Molden, D., U. Amarasinghe, and I. Hussain. 2001. "Water for Rural Development: Background Paper on Water for Rural Development Prepared for the World Bank.". Working Paper 32. IWMI, Colombo, Sri Lanka.

Moore, D., and D. Seckler. 1999. *Water Scarcity in Developing Countries. Reconciling Development and Environmental Protection.* Arlington, VA: Winrock International Institute for Agricultural Development.

National Research Council. 1996. *A New Era for Irrigation*. Washington, DC: The National Academies Press.

OECD (Organisation for Economic Co-operation and Development). 2004. Issues note for the second workshop of the infrastructure for poverty reduction task team. Berlin.

Oweis, T., D. Prinz, and A. Hachum. 2001. *Water Harvesting: Indigenous Knowledge for the Future of the Drier Environment.* Aleppo, Syria: ICARDA.

Oyebande, L. 2004. "Fostering Implementation: Know-How for Action." Annex to the Report of the African Pre-conference "Water for Food and Ecosystems: Make it Happen!" Addis Ababa, November 4–6.

Pimentel, D., B. Berger, D. Filiberto, M. Newton, B. Wolfe, E. Karabinakis, S. Clark, E. Poon, E. Abbett, and S. Nandagopal. 2004. "Water Resources: Agricultural and Environmental Issues." *BioScience* 54 (10): 909–18.

Plusquellec, H. 1997. "The Future of Irrigation." Presented at the Regional Seminar on Sustainable Irrigation, Tashkent, February 20–25.

———. 2002. *How Design, Management and Policy Affect the Performance of Irrigation Projects. Emerging Modernization Procedures and Design Standards.* Bangkok, Thailand: FAO.

Postel, S. 1999. *Pillar of Sand: Can the Irrigation Miracle Last?* New York, London: W. W. Norton & Company.

Priscoli, J. D. 2005. "Water Institutional Reforms: Theory and Practice." *Water Policy* 7 (1): 1–19.

Renault, D., W. W. Wallender. 2000. "Nutritional Water Productivity and Diets." *Agricultural Water Management* 45 (3): 275–96.

Roe, T., A. Dinar, Y. Tsur, and X. Diao. 2004. "Feedback Links between Economy-Wide and Farm-Level Policies: Application to Irrigation Water Management in Morocco." Policy Research Working Paper No. 3550, World Bank, Washington, DC.

Rosegrant, M. W., and M. Svendsen. 1993. "Asian Food Production in the 1990s: Irrigation Investment and Management Policy." IFPRI, Washington, DC.

Rosegrant, M. W., X. Cai, and S. A. Cline. 2002a. *Global Water Outlook to 2025: Averting an Impending Crisis.* Washington, DC: IFPRI/IWMI.

———. 2002b. *World Water and Food to 2025: Dealing with Scarcity.* Washington, DC: IFPRI.

Schultz, B. 2002. Opening Address. ICID's 18th International Congress on Irrigation and Drainage. Montreal, Canada, July 21–28.

Shah, T., B. van Koppen, D. Merrey, M. de Lange, and M. Samad. 2002. *Institutional Alternatives in African Smallholder Irrigation: Lessons from International Experience with Irrigation Management Transfer.* Research Report 60. Colombo, Sri Lanka: IWMI.

Showers, Kate B. 2002. "Water Scarcity and Urban Africa: An Overview of Urban-Rural Water Linkages." *World Development* 30 (4): 621–48.

SIWI (Stockholm International Water Institute). 2004. *World Water Week: Report on the Seminar on Financing Water Infrastructure.* Stockholm.

SIWI-IWMI (Stockholm International Water Institute–International Water Management Institute). 2004. *Water – More Nutrition per Drop. Towards Sustainable Food Production and Consumption Patterns in a Rapidly Changing World.* Stockholm.

Strzepek, K., D. Molden, and H. Halbraith. 2001. "Comprehensive Global Assessment of Costs, Benefits and Future Directions of Irrigated Agriculture: A Proposed Methodology to Carry Out a Definitive and Authoritative Analysis of Performance, Impacts and Costs of Irrigated Agriculture." Dialogue Working Paper 3. Dialogue Secretariat. IWMI, Colombo, Sri Lanka.

Sur, M., D. Umali-Deininger, and A. Dinar. 2002. "Water-Related Subsidies in Agriculture: Environmental and Equity Consequences." OECD Workshop on Environmentally Harmful Subsidies, Paris, November 7–8.

Svendsen, M., J. Trava, and S. H. Johnson III. "Participatory Irrigation Management: Benefits and Second Generation Problems." Lessons from an International Workshop held at Centro Internacional de Agricultura Tropical (CIAT). Cali, Colombia, February 9–15.

Swedish Water House and the Millennium Project. 2005. "Investing in the Future. Water's Role in Achieving the Millennium Development Goals." Swedish Water House Policy Briefs NR1.

Tardieu, H., and B. Prefol. 2002. "Full Cost or 'Sustainability Cost' Pricing in Irrigated Agriculture: Charging for Water Can Be Effective, But Is It Sufficient?" *Irrigation and Drainage* 51 (2): 97–107.

Tsur, Y., and A. Dinar. 2004. *Pricing Irrigation Water: Principles and Cases from Developing Countries.* Washington, DC: Resources for the Future.

United Nations. 2003. *Sectoral Water Allocation Policies in Selected ESCWA Member Countries:. An Evaluation of the Economic, Social and Drought-Related Impact.* New York: UN Economic and Social Commission for Western Asia.

———. 2004. Millennium Project Hunger Task Force. "Halving Hunger: It Can Be Done. An Action Plan for Implementing the Millennium Development Goal on Hunger." Draft Report. UNDP, New York.

Upadhyay, B. 2003. "Water, Poverty and Gender: Review of Evidences from Nepal, India and South Africa." *Water Policy* 5 (5-6): 503–11.

Van Hofwegen, P., and M. Svendsen. 2000. "A Vision of Water for Food and Rural Development." Sector Report. World Water Council, Marseille, France.

Van Koppen, B., S. Mahmud, and W. Nicholaichuk. 1996. "Private Irrigation by Poor Women and Men in Bangladesh: Institutional Issues." In *Sustainability of Irrigated Agriculture Farmer's Participation toward Sustainable Irrigated Agriculture*, volume 1B. Transactions of the Sixteenth Congress on Irrigated Agriculture. Cairo: ICID.

Van Koppen, B., J. van Etten, P. Bajracharya,. and A. Tuladhar. 2001. "Women Irrigators and Leaders in the West Gandak Scheme, Nepal." Working Paper 15. IWMI, Colombo, Sri Lanka.

Van Steenbergen, F. 2002. *Local Groundwater Regulation.* Water Praxis Document Number 14. Land and Water Product Management Group. Arnhem, The Netherlands: Arcadis Euroconsult.

Van Vuren, G., and A. Mastenbroek. 2000. "Management Types in Irrigation: A World-Wide Inventory per Country." Report commissioned by the World Bank. CWP research paper 1, Wageningen University, Wageningen, The Netherlands.

Vermillion, D. L. 1997. *Impacts of Irrigation Management Transfer: A Review of the Evidence.* Research Report 11. Colombo, Sri Lanka: IWMI.

———. 2004. "Taking Action for Participatory Irrigation Management." Draft background report for *Directions in Development. Reengaging in Agricultural Water Management: Challenges, Opportunities, and Trade-Offs.* Washington, DC: World Bank.

Vermillion, D. L., and J. Sagardoy. 1999. *Transfer of Irrigation Management Services: Guidelines.* Irrigation and Drainage Paper 58. Rome: FAO.

Vidal, A., C. Rigourd, and A. N. de Villemarceau. 2004. *Identification et diffusion de bonnes pratiques sur les perimeters irrigues en Afrique de L'Ouest.*

Rapport technique final. Programme international pour la technologie et la recherché en irrigation et drainage. Rome: IPTRID.

Winpenny, J. 2003. *Financing Water for All.* Report of the World Panel on Financing Water Infrastructure. World Water Council, 3rd World Water Forum, Global Water Partnership.

———. 2005. "Financing Water for Agriculture." An issues paper in preparation for the Fourth World Water Forum, Mexico 2006.

World Bank. 1994. *A Review of World Bank Experience in Irrigation.* Report No. 13676, Operations Evaluation Department, Washington, DC.

———. 2000a. "Nigeria. National Fadama Development Project." Implementation Completion Report No. 19730. Rural Development 2, Africa Regional Office.

———. 2000b. "Pricing Irrigation Water. A Literature Survey." Policy Research Working Paper 2449, Rural Development Department, Washington, DC.

———. 2002a. *Bridging Troubled Waters: Assessing the World Bank Water Resources Strategy.* Operations Evaluation Department. Washington, DC.

———. 2002b. "China Country Water Resources Assistance Strategy." East Asia and Pacific Region.

———. 2002c. *A Sourcebook for Poverty Reduction Strategies.* Jeni Klugman, ed. Washington, DC.

———. 2003. "Prospects for Irrigated Agriculture: Whether Irrigated Area and Irrigation Water Must Increase to Meet Food Needs in the Future. Validation of Global Irrigation-Water-Demand Projects by FAO, IFPRI, and IWMI." Report No.26029, Agriculture and Rural Development Department, Washington, DC.

———. 2004a. "Brazil. Irrigated Agriculture in the Brazilian Semi-Arid Region: Social Impacts and Externalities." Report No. 28785-BR. Washington, DC.

———. 2004b. "Drainage for Gain: Integrated Solutions to Drainage in Land and Water Management." Agriculture and Rural Development Department. Washington, DC.

———. 2004c. "PPP for Irrigation and Drainage. Need for a Professional 'Third Party' Between Farmers and Governments." Background report for *Directions in Development. Reengaging in Agricultural Water Management: Challenges, Opportunities, and Trade-Offs.* Washington, DC.

———. 2004d. "Iran Country Water Resources Assistance Strategy." Washington, DC.

———. 2004e. *Water Resources Sector Strategy: Strategic Directions for World Bank Engagement.* Washington, DC.

———. 2005a. *Agricultural Growth for the Poor: An Agenda for Development.* Directions in Development series. Washington, DC: World Bank

———. 2005b. *Shaping the Future of Water for Agriculture: A Sourcebook for Investment in Agricultural Water Management*. Washington, DC.

———. Forthcoming. "Managing Water Resources to Maximize Sustainable Growth: A Country Water Resources Assistance Strategy for Ethiopia." Washington, DC.

World Bank and Stockholm International Water Institute. 2004. Report on the Seminar on Financing Water Infrastructure. World Water Week, Stockholm.

World Health Organization. 2000. "Gender, Health and Poverty." Fact Sheet no. 25, June.

Zimmer, D., and D. Renault. 2003. "Virtual Water in Food Production and Trade at Global Scale: Review of Methodological Issues and Preliminary Results." Proceedings of Expert Meeting on Virtual Water, Delft, December 2002.

Zimmer D., M. Mermond, M. L. Melliand, and S. Tato-Serrano. 2002. *Global Drainage Status and Needs*. Antony Cedex, France: CEMAGREF.

Zwart, S. J., and W. G. M. Bastiaanssen. 2004. "Review of Measured Crop Water Productivity Values for Irrigated Wheat, Rice, Cotton and Maize." *Agricultural Water Management* 69 (2): 115–33.

Zwarteveen, M. Z., and N. Neupane. 1996. *Free-Riders or Victims: Women's Nonparticipation in Irrigation Management in Nepal's Chhattis Mauja Irrigation Scheme*. Research Report 7. Colombo, Sri Lanka: IWMI.

Index